Springer Geography

The Springer Geography series seeks to publish a broad portfolio of scientific books, aiming at researchers, students, and everyone interested in geographical research. The series includes peer-reviewed monographs, edited volumes, text-books, and conference proceedings. It covers the entire research area of geography including, but not limited to, Economic Geography, Physical Geography, Quantitative Geography, and Regional/Urban Planning.

More information about this series at http://www.springer.com/series/10180

Anze Chen · Yunting Lu · Young C.Y. Ng

The Principles of Geotourism

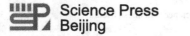 Science Press
Beijing

 Springer

Anze Chen
Chinese Academy of Geological Sciences
Beijing
China

Young C.Y. Ng
Association for Geoconservation,
   Hong Kong
Hong Kong
China

Yunting Lu
School of Geography
Beijing Normal University
Beijing
China

ISSN 2194-315X   ISSN 2194-3168   (electronic)
Springer Geography
ISBN 978-3-662-52617-0   ISBN 978-3-662-46697-1   (eBook)
DOI 10.1007/978-3-662-46697-1

Jointly published with Science Press, Beijing
ISBN: 978-7-03-043495-1 Science Press, Beijing

Springer Heidelberg New York Dordrecht London
© Springer-Verlag Berlin Heidelberg and Science Press Ltd. 2015
Softcover reprint of the hardcover 1st edition 2015

Printed on acid-free paper

Springer-Verlag GmbH Berlin Heidelberg is part of Springer Science+Business Media
(www.springer.com)

Foreword 1

The Principles of Geotourism (1991), compiled by Anze Chen, Yunting Lu, et al. of China Tourism Earth-science Research Association, is a summary of research findings of several 100 members of the Association in five years since its establishment, as well as a pioneering move of China's earth-science workers to serve tourism and apply the theories and methods of earth-science to tourism. It basically reflects China's research level of contemporary tourism earth-science. The publication of this work is of significance for both the earth-science circle and the tourist circle. I, as the honorary chairman of the Association, feel very happy about so many research findings in just several years after the establishment of the Association, and feel it necessary for me to recommend this monograph to the earth-science, tourist and academic circles and say a few words about the birth of this new discipline.

"Tourism earth-science" is an emerging marginal discipline which uses the theories and methods of earth-science and draws the essences of other disciplines to serve the needs of survey, evaluation, planning, development, management and protection of tourism resources, thereby boosting tourism development. It comes into being along with the emergence of China's tourism industry. It is known to all that tourism, as a social phenomenon, has been in existence in China for thousands of years. However, tourism as an industry did not take shape in China until 10 years ago. In the late 1970s, China launched its reform and opening up, giving rise to the booming development of tourism, and in the early 1980s, tourism has come up as a sizeable sector of national economy. In this process, varied industries and disciplines had been penetrating into tourism and serving tourism by various scientific means. Geological and geographical circles take the lead. They have not only inaugurated "tourism earth-science", but also achieved a series of highly academic research findings, built up a research team with adequate qualifications, thereby promoting the formation and development of tourism earth-science.

Despite different research areas, geology and geography belong to the same category of Earth-science and both serve the tourism industry. With a view to finding an intersection of the two in tourism researches, a seminar on tourism earth-science was held in the spring of 1985 as proposed by Committee of

Science Popularization of Geological Society of China. Scholars of geological, geographical, gardening, environmental and tourist circles got together, exchanging their research findings in combining with tourism science and serving tourism. All the attendants felt much wiser and unanimously thought it necessary to set up an academic team beyond all branches of science in order to research and explore the varied earthscientific issues relating to tourism. On careful discussion and consultation, the attendants accepted "tourism earth-science" as the name of the new discipline, and "China Tourism Earth-science Research Association" was established. Comrade Daguang Sun, the then Minister of Geology and Mineral Resources, Prof. Renzhi Hou, a famous geographer, and I (the then chairman of Geological Society of China) were elected honorary chairmen. I was pleased to take this honorable position. After establishment of the Association, the members have enriched and improved theories and gradually built up this new discipline—tourism earth-science in line with the guidelines of science and technology serving the construction of national economy and "practice–theory–further practice".

As assigned by State Planning Commission and Bureau of Land and Resources in the year of establishment of the "Association", it drew the perspective zoning map (1:6,000,000) of China's tourism resources, showing over 70 tourist areas and spots, and proposed the preliminary scheme of dividing national scenic areas into 86 sites (zones), which basically meets the requirement of the state for tourism resources in formulating national land rectification strategies. In recent years, the Association focused on researches on tourism resources and their development principles, including researches on structures, combinations, types and distribution rules of tourism resources; made drawings of various national and regional tourism resources, and provided plenty of basic materials for central and local governments to develop tourism; researched the formation conditions and evolution models of tourism resources, enriching the formation theories of tourism resources; drew perspective forecast maps of tourism resources to point out the direction for exploring new tourism resources; studied the methods of survey and evaluation of tourism resources, and development planning principles, reasonable use and protection schemes of tourist areas (spots) to provide scientific and technological certifications for the overall development of tourism resources; studied earth-science and other sciences like history, economics, psychology and aesthetics relating to tourism; and studied the popularisation of earth-science in tourism, and combined tourism with spiritual civilisation education as well as the improvement of the scientific and cultural level of the whole nation. In a word, the studies on the above-mentioned issues have enriched and deepened the contents of tourism earth-science from different scopes and perspectives.

Tourism earth-science, as an emerging science, needs more of our attention and support so as to give greater play to its potentials and functions in the expansive tourism field. However, we should also note that the theories, methods and contents of an emerging discipline are all pending for enrichment, improvement, as well as varied ordeals of practices. As we enter into the twenty-first century, I feel deeply that earth-science needs to develop and have its connotation deepened and service sectors expanded, and create more new disciplines with endless vitality like tourism earth-science during practices. I believe, by virtue of the concerted efforts of earth-science and tourist circles, "tourism earth-science" aiming to serve tourism will surely have its theoretical and method systems improved day by day, and fully display its irreplaceable effects and values in China's tourism.

December 1989   Jiqing Huang

Foreword 2

Earth, is a small rocky planet hurtling through space. From the edges of the solar system, it is barely visible as a small, blue dot. But this small, blue dot is special. Very special. For 4,600,000,000 years, as it orbits the Sun, the surface of this small planet has altered innumerable times. As the tectonic plates of the Earth's thin solid outermost layer have moved across the eons of time, so oceans have come and gone, continents have collided and split apart, mountain ranges have risen and then been worn away and life, in all of its incredible diversity, has evolved and multiplied. How do we know all this? We know because the memory of these things are written in the stones and rocks all around us. Professor Anze Chen is one of those very special people who can not only read the memory of the Earth as it is recorded in the rocks, but also has the ability to share these stories with humanity, re-connecting them to the planet we call home. By engaging the Chinese people with their geological heritage, with their rich geodiversity, Prof. Chen has also assisted in the development of geotourism in China, and through that, in the sustainable economic development of communities across China. Through the sustainable approach used in geotourism and adopted by Global Geoparks, Prof. Chen has helped ensure that the geological heritage of China will remain intact for future generations to enjoy and that, through this English version of his book, the amazing journey of our small planet becomes more widely known across the world.

Prof. Patrick J. Mc Keever
Chief of Section, Earth-sciences and Geohazard Risk Reduction
Secretary of the IGCP
EES/NGR
Natural Sciences Sector/Bureau 5.08
UNESCO

Foreword 3

We all love our nature and stand ready to protect it, especially against the increasing human and urban encroachment. Rocks, landscapes and geological formations are not just spectacular creations by our mother nature but also the bedrock of our ecosystem and living planet. The understanding of the making of our planet earth should not be confined to academic pursuit. If we make it simple and well articulated, it could foster a greater sense of ownership by all walks of life and encourage all communities in the world to preserve our nature. Professor Anze Chen is a forerunner in this endeavour and his deeds, including the publication of this new book, set a good example.

For decades, Prof. Anze Chen has pioneered the concepts of tourism earth-science and geopark in China. These concepts, supported by practical works and demonstrated results, have convinced geoscience academia and policy makers to promote the establishment of national geoparks and later UNESCO's accredited global geoparks in China including the one recently established in Hongkong. Apart from effectively protecting China's invaluable geological heritage and scenic wonders, well-managed geoparks have proven to be a healthy and sustainable engine bringing about improvement to the livelihood of the local people especially those in remote and less developed areas. On the education front, green-tourism-cum-earth-science, as a multidisciplinary academic subject, has gained its place and popularity among schools and colleges particularly for students studying geology and geography as well as those pursuing studies in tourism and development.

I got to know Prof. Chen during the establishment of the Hongkong Geopark. We share the same vision of geoconservation and I am deeply impressed by his dedication and relentless efforts in promoting tourism earth-science in China for the benefits of all people. Professor Chen and his associates published the original Chinese version of their ground-breaking book 'The Principles of Tourism Earth-science' in 1989. They now take a step forward by publishing this English version to promote their ideas in the international arena. I take this opportunity to wish him and his team every success in this meaningful and challenging endeavour.

August 2014   Edward Yau Tang-wah

Edward Yau, during his tenure as Hong Kong's Secretary for the Environment (2007–2012), championed the establishment of the Hong Kong Geopark (2009), which later became UNESCO's accredited global geopark (2011). Mr. Yau is currently the Director of the Chief Executive's Office of the Hong Kong Special Administrative Region.

Preface

Geotourism is an alternative tourism which may be categorised under nature tourism or alternative tourism. It can be practiced in both natural and urbanised areas. Its main focus is the sustainable use of geology and natural landscape as tourism resources to attract tourists, provide geoscience knowledge to the public and students and encourage appreciation and develop a sense of place and value for protection. The National Geographic has a different definition as it views geotourism as geographical tourism which embraces geographical characteristics as tourist attraction. Geographical characteristics include varied natural and cultural features of a place and do not highlight the values of geology and natural landscape which form the foundation of all ecological and cultural components of a tourist destination. This definition makes not much distinction when compared to the usual mass tourism.

Geotourism in this book has been interpreted as geological tourism with focus on geology and its interaction with ecology and culture. It is since the early 1980s that this geotourism has been named 'tourism earth-science' in China. Chinese geologists have been adopting geoscientific methods in exploring tourism opportunities provided by geology and the natural landscape. This book was first published in Chinese in 1991 by Peking University Press under the title of *An Introduction to Tourism Earth-science*. It is the first volume of *China's Geotourism Series,* compiled under the auspices of the Chinese Academy of Tourism earth-science and Geopark Research, Geological Society of China. It summarises the Academy's research findings in geotourism in the past few decades with the objective to enhance the theoretical level of this emerging multidisciplinary subject and to improve its research methods for its long-term development. Such development has far-reaching impact on China's tourism due to the country's extremely rich geodiversity. In retrospect, the decision to publish the Chinese version was made during a work meeting held in Wuhan in 1986 when some members of the Academy drafted the outline of a book to encourage the application of earth-science in tourism. In 1987, the Academy invited its members to write academic papers on theories and methodology of this new perspective of applied geoscience. It has since then been described in Chinese as 'tourism earth-science'.

During the seminar in Mount Tianmu on tourism earth-science, the content of the book was further dicussed, amended and enhanced with additional contribution of the research findings of members of the Academy. An editorial board of *An Introduction to Tourism Earth-science* was set up after the seminar to take care of the contribution of chapters. It took another year before the book was finally completed. This book was therefore a collective work of enthusiastic members whom I would like to extend my greatest gratitude to.

The idea of having an English version of *An Introduction to Tourism Earth-science* stems from the rapid growth of global geoparks and geotourism around the globe since 2004. The English version of this book is renamed *The Principles of Geotourism* to provide global readers a definition and theoretical framework of geotourism, which has used and practised in China for over three decades. This book originally contained 11 chapters: Chap. 1 by Anze Chen and Yunting Lu; Chap. 2 by Yunting Lu; Chap. 3 by Erkuang Zhang; Chap. 4 by Qinglian Wang; Chap. 5 by Kang Guo; Chap. 6 by Ninggao Xie; Chap. 7 by Weixin Li and Yunting Lu; Chap. 8 by Yunting Lu and Chap. 9 by Daolong Xing. Two new chapters Chap. 10 on Geopark and Chap. 11 on Prospect of Tourism Earth-science were added by Anze Chen to update the latest development in geoparks and their impact on geotourism. The late Prof. Yunting Lu and I were editors of the original Chinese version while Guangzhen Cui had made final revisions to the book. The editorial board was later joined by Dr. Young Ng who had prepared and edited the English version of the book. On behalf of the editorial board, we welcome any suggestions and comments which will definitely be very helpful for us to prepare for any future modification of the book.

Beijing, China, October 2014   Prof. Anze Chen
President, Academy of Tourism Earth-science
and Geopark Research, the Geological Society of China

Contents

List of Contributors

China Tourism Earth-science Research Association

Editors-in-chief

Anze Chen
Yunting Lu
Young C.Y. Ng

Editorial Board Members

Weixin Li
Erkuang Zhang
Qinglian Wang
Kang Guo
Ninggao Xie
Daolong Xing

Chapter 1
Research Objects, Tasks and Historical Development of Tourism Earth-science

1.1 Research Objects of Tourism Earth-science

1.1.1 Definition of Tourism Earth-science

Every science has its unique research objects. Tourism earth-science is no exception. Its research objects differ from those of geology, geography and tourism science, but also interact as both cause and effect with those of the aforesaid disciplines due to the characteristics of it as a marginal discipline. Despite insufficient understanding about this emerging discipline, its objective applications and research value have drawn extensive attention. Tourism earth-science, which has objective guidance functions and grows gradually during scientific practices, will surely find extensive applications along with the development of tourism. Therefore, what is the most important for us in studying tourism earth-science is to make clear the nature and research objects of tourism earth-science.

What is tourism earth-science? To answer this question, we may first make an overview of several scientific viewpoints of tourism earth-science.

"Proposition of tourism geology" This is a viewpoint about earth-science in its narrow sense. The scholars holding this viewpoint limit tourism earth-science to the research domain of tourism geology and think that tourism earth-science is a science which studies the geological and geomorphic conditions of tourist spots (areas). This viewpoint focuses on making tourism geological surveys for serving professional geological travels and spreading geological knowledge to travellers. Such an understanding generally appeared in the early stage of China's tourism. At that time, scholars of all disciplines strived to serve tourism, but they knew little about tourism as a whole and were not sure how to serve tourism. Therefore, what they did was just to integrate tourism into the discipline that they were engaged in. The geological circle compiled and published *Geological Series on China's*

© Springer-Verlag Berlin Heidelberg and Science Press Ltd. 2015
A. Chen et al., *The Principles of Geotourism*, Springer Geography,
DOI 10.1007/978-3-662-46697-1_1

Places of Interest. It includes some guidance materials for some professional geological survey routes or some popular guidance materials, which were regarded as the contents of tourism geology. At that time, tourism geology was also referred to as "scenery geology", "landscape geology", "garden geology", "geology of places of interest", etc. Later, this view developed to include exploring and evaluating tourism geological resources as research objects of tourism geology in a bid to provide tourism resources. Today, many people still equate tourism earth-science with tourism geology.

"Proposition of tourism geography" Liu Xingshi wrote in the article entitled "On Tourism Earth-science", "Tourism earth-science is a special geography which serves tourism". He argued that tourism earth-science focused on research on natural sceneries themselves, fell basically within the domain of physical geography and was also involved in human geography and other social sciences. He held that tourism earth-science not only directly served travellers through providing scientific guidance, spreading scientific knowledge, introducing places of interests and so on, but also served tourism in such respects as research and development of tourism bases, national land rectification and environmental protection. This proposition expanded the research domain as compared to the preceding one, but it still failed to summarize the research domain of tourism earth-science as a whole. Meanwhile, it is also inappropriate to define tourism earth-science as a special geography.

"Proposition of tourism earth-science" This proposition treats tourism earth-science in its broad sense. It argues that tourism earth-science is a marginal discipline born from the marriage of earth-science and tourism science, which mainly includes two parts: tourism geology and tourism geography. The research contents of tourism earth-science should cover all earth scientific issues of tourism or all tourism issues studied by earthscientific theories and methods. Tourism earth-science should study not only the object of tourism—tourism resources—but also two subjects of tourism—tourism markets and media as well as tourism services and facilities. In a word, all the three elements of tourism have earthscientific topics and rules pending to be explored by earthscientific workers. Chen Anze and Li Weixing put forward in the article entitled "Status Quo and Prospects of Tourism Earth-science", "Tourism earth-science is a comprehensive marginal discipline which is aimed to find, evaluate, plan and protect natural landscapes and cultural relics with tourism value, and discuss their formation causes and evolution history on the basis of earthscientific theories and methods and in combination with the knowledge of other disciplines, with a view to promoting the development of tourism".[1] Zhu Liangpu stated in the article entitled "Suggestions and Discussion on Work Methods of Earthscientific Classification of Tourist Spots", "Tourism earth-science takes tourism related earthscientific issues as research objects and the development of tourism as research purpose". Zhang Erkuang et al. wrote in the

---

[1] Chen Anze, Li Weixing, 1985, Status Quo and Prospects of Tourism Earth-science, *The Earth*, (4–16).

article entitled "Preliminary Discussion on Basic Characteristics and Development and Utilization of Tourism Earthscientific Resources in Hebei Province", "Tourism earth-science is an applied science which serves the survey, planning and development of tourist spots by modern earthscientific theories and scientific and technological means, and is aimed to promote academic exchanges on earth-science and spread earthscientific knowledge". Cui Guangzhen elaborated the definition of tourism earth-science as follows in the article entitled "On China's Study of Tourism Earth-science", "Tourism earth-science, as a new applied science, emerges to meet the need of tourism development, namely the application of earth-science in tourism; tourism earth-science is born from the marriage of earth-science and tourism science and forms an important branch of earth-science, and it includes two domains—tourism geology and tourism geography, so it is also a comprehensive marginal science" and "Its research object is the objective tourism environments, including all appreciation contents relating to earth-science". (Cui 1989) Therefore, this viewpoint has strongly reflected the idea that basic earthscientific theories and methods should be applied to research various issues relating to tourism science, in order to serve tourism.

To sum up, tourism earth-science is understood in the earthscientific circle either in the broad sense or in the narrow sense. The former stresses the research on all earthscientific issues in tourism by earthscientific theories and methods, while the latter emphasizes geology or geography. We agree to the proposition of the broad sense of tourism earth-science.

With a view of clearly defining tourism earth-science, we think it is necessary to follow the three principles: (1) persisting in combining earth-science with tourism science and tourism; (2) persisting in combining theory with practice; and (3) persisting in combining tourism geology with tourism geography. These three "combinations" are key to fully and accurately defining tourism earth-science.

Earth-science includes such disciplines as geology, geography, geochemistry, geophysics and atmospheric physics. In particular, geology and geography are the main branches of earth-science. The former studies the composition, structure, conformation and development history of earth, while the latter studies earth surface, namely the geographical environment where human beings live. The research contents of the two, regardless of the components of the earth, various energies and dynamic forces in the components, varied geological structures formed under the energies and dynamic forces, and rocks and mineral distributions in the crust, as well as various geographical elements constituting the earth surface—landforms, climate, hydrological conditions, vegetation, soil, production activities, residents, etc., are all earthscientific bases of the formation of tourism resources. That is to say, the components of the earth and spatial distributions and changes of energies lead to the diversification, multi-layers and polytropism of landscapes on the earth surface. Such landscapes on the earth surface include natural landscapes and cultural landscapes, which man uses for sightseeing, rest, rehabilitation and recreation. To meet and improve such a relationship between man and sceneries, people begin to study a special science on the relationship between earth-science and tourism—tourism earth-science. Therefore, tourism earth-science, as an emerging branch of earth-science, studies the relationships between sightseeing, recuperative and recreational activities of human beings, and components and structures of earth surface as well as transfer and changes of energies. In other words, tourism earth-science studies the laws of the relationship between tourism and earthscientific environments. It covers geological and geographical tourism environments. Therefore, tourism earth-science is a generic term for two marginal sciences—tourism geology and tourism geography.

1.1.2 Definition of Tourism Geology

Tourism geology is a branch of tourism earth-science. It studies the distributions, types, characteristics, causes of formation and changes of varied scenic spots by geological theories, methods, technologies and results. The comprehensive survey and evaluation of basic geology, karst geology, dynamic geology and environmental geology of scenic areas and spots are conducted to organize targeted earthscientific travels, reasonably select travel routes and supporting facilities, maximally display the aesthetic, cultural and scientific values of scenic areas and spots, and integrate scientificity and interest, which constitute an emerging interdisciplinary science with unique research objects and methods. Therefore, tourism geology is a highly applicable and comprehensive marginal science, which combines geology and tourism, and is also an emerging interdisciplinary science which is expanded and extended from the perspective of actual needs of tourism. Its birth indicates the extensive and close relations between the two sciences in the following four aspects: (1) common research and development objects; (2) identical objectives; (3) borrowable methods and technologies; and (4) inevitable combination of their development trends.

These relations between geology and tourism science may be evidenced in many aspects. Natural landscapes including peaks, valleys, caves, stones, rivers, lakes, waterfalls and springs are all generated by geological processes and controlled by varied geological factors. Complicated geological factors lead to diversified shapes of natural landscapes. In particular, structural features generated by tectonic activities, and their natures, characteristics, scales and attitudes all directly control the formation and development of natural landscapes. From the macroperspective, they can control the pattern of geomorphic units, shapes, trends and elevations of mountains, layout of water systems, and formation of rivers, lakes and groundwater; from the micro-perspective, they control the development of peaks, valleys, caves and springs. For example, Mount Putuo, Zhejiang, which has been swelling in geological structure, is a part of Cathaysia. Blocks with violent Yanshan movement formed a series of grabens and horsts, which have laid a foundation for the geomorphic development of Zhoushan Islands. In the Late Kainozoic era, Mount Putuo had been submerged by the sea. However, new tectonic movements later raised it sporadically, displaying it as an island with varied beautiful marine abrasion and deposition landforms and mountain and sea blended scenes. For another example, the formation of springs in Jinan, known as City of Springs, is mainly ascribable to its geological structure. Jinan is located on the north slope of the domal structure of Mount Taishan with the gradient of 10°–15° and has faults, such as Thousand-Buddha Mountain Fault, which facilitate the confluence, run-off and drainage of groundwater. Volcanic activities and their resulting unique landscapes have more significant effects on creating earthscientific tourism values. Beautiful sights such as the stone ridges, stone sea, flowerlike rocks, lava tunnels and lava waterfalls of Wudalianchi Lakes and Jing po Lake are all directly related to activities of lava flows during volcanic eruption.

Moreover, strata and rocks are also the most common and most important geological factors that affect natural landscapes. Different strata have different rock formations, rocks, attitudes, thicknesses and temporal and spatial distribution rules, and different rocks have different mineral compositions, chemical compositions, structures, tectonics, colours, thicknesses, specific gravity, permeability, and other chemical and physical properties. All these factors geologically brand natural landscapes. For example, with grand, arduous, deep and simple landscapes, Mount Huangshan and Mount Hua composed of granites look like a vivid landscape painting either in close quarters or from a distance and have been well known as "Mount Huangshan the most wonderful under heaven" and "Mount Hua the most arduous under heaven"; landscapes such as Guilin scenery, Three Gorges on the Yangtze River and Stone Forest of Lunan composed of limestones and formed under the impact of tropical and subtropical lava are delicate and pretty, easy, deep and peculiar, look like exquisite traditional Chinese realistic paintings, and are famous in the world for their four wonders—"mountain, water, cave and stone"; scenic spots composed of red sandstones and conglomerates, such as Mount Danxia, Mount Wuyi and Mount Maiji, are arduous, bold, gorgeous and torrid, and look like elaborate sculptures from any perspectives. Therefore, all natural landscapes are formed under specific geological conditions. Based on this internal link, tourism geology studies the distributions, shapes, natures, characteristics, types and development factors of various natural landscapes by geological theories and methods, which have laid a foundation for the establishment of it as a branch of tourism earth-science.

The extensive and close relationship between tourism and geology is not limited to natural landscapes, because geological contents are also found in many research topics of cultural landscapes. For example, the selection and processing of varied stone materials used in some historical buildings, commemorative archways, monuments, towers, columns, as well as landscaping pavilions, bridges, corridors, and gardens depended on outcrops, geological conditions, and physical and chemical properties of different rocks and were therefore closely related to technological petrology and engineering geology of applied geology; besides, in the vicissitudes of thousands of years, some cultural relics and historical sites recorded and stored the traces of some geological processes, bearing witness to crustal movements and geological processes. Moreover, the site of ancient Handan City from the Warring States Period (475 B.C.–221 B.C.) to the Han Dynasty (202 B.C.–220 A.D.) was cut by a north-northeast active fault. Its eastern part sank and was buried 7–9 m under the earth surface, while the western part was uplifted, exposed on the earth surface and cut by gulleys. Another example is the offset and displacement of the Great Wall at Huangyaguan, the settlement of the iron lion at the historical site of Kaiyuan Temple, Jiuzhou (the ancient Cangzhou) of the Song Dynasty (960 A.D.–1279 A.D.), the connection of Jinshan (Golden Hill) Pagoda of Zhenjiang and the south bank of Yangtze River, and the incline of Tiger Hill Pagoda of Suzhou, etc., which are all bound up with geological processes.

1.1.3 Definition of Tourism Geography

Tourism geography is also a branch of tourism earth-science. It has been developing with tourism science against the background of scale development of tourism. Tourism science is a branch of social economics with tourism as its research object and falls under the category of social sciences. However, its research covers a lot of geographic problems including massive basic theoretical knowledge of natural geography and human geography. That is to say, tourism science should be based on certain theoretical knowledge of geography, while the research on the geographical problems involved in tourism is a direct application of geography. Therefore, tourism geography is a natural result when modern tourism has developed to a certain stage. Elina Amblovich, a famous professor of Soviet Union, points out in *Tourism Geography*, "Tourism phenomenon happens on various occasions of natural geography and human geography, and is therefore most directly linked to geography. As a result, a specialized geography, namely the young tourism geography, has emerged to study tourism".

Tourism geography is not only a branch of the emerging earth-science (Lu 1988), but also a comprehensive interdisciplinary science. Since tourism geography studies the various geographic problems connected with tourism, it naturally has the marginal characteristics of tourism science and geography. However, since geography has so many branches, views disagree on the margin of the branches. Some scholars regard tourism geography as a branch of economic geography, some regard it as a branch of human geography and some others regard it as a branch of cultural geography. The comprehensiveness of tourism geography is mainly reflected by the universal and interdisciplinary nature of its research domain. Apart from all the branches of geography, tourism geography also involves sociology, folkloristics, archaeology, history, architecture, garden architecture, environmentology, economics, ethnology, literature, aesthetics, psychology, etc. Tourism geography is also a practical science. On the one hand, it has its own theoretical system for researches on generation and development rules, characteristics and types of tourism resources; on spatial distribution and trend of spatial movement of travellers; on development, protection, layout and planning of tourist sites; and on the influence of the development of tourism on the formation of regional economic complexes. On the other hand, it also stresses research on utility, especially in countries with developed tourism in the world. For example, some European and American countries have been treating tourism geography as a highly practical science, which studies extensive applied projects, such as utilization and planning of recreational land; opening and estimation of capacity of national parks, state parks and other tourism sites; the discovery and orientation of resorts; and measurement technique and forecast technique of tourist flow. Many countries even directly take tourism geography as the basic teaching materials for training tourism professionals and providing guides and various talents for tourism development and management.

In recent years, the scholars of China have been understanding tourism geography in increasing depth. In the early 1980s, some scholars held that "Tourism geography studies the distribution rules, development conditions and utility value of tourism resources, planning and layout of tourist areas, spots and lines, as well as the constraint relationship with the ambient economic complexes".[2] Later, by summing up tourist practices, they further understood that "Tourism geography should not be limited to studying the development of tourism resources and tourist sites, but should also make comprehensive researches on the geographical problems relating to tourism development based on the three key elements of tourism, and comprehensively discuss the rules of formation, evolvement and development of these problems". For example, it studies the concepts, classification, distribution rules, survey, assessment, appreciation and development of tourism resources, reasonable distribution and planning of tourism sites; the relationship between tourist activities and facilities and geographical environment; tourism motives, directions of tourist flow and changes in quantity and quality of tourism environments; and traffic lines for travels, design of tourism maps and tourism zoning. However, the research objects of tourism geography still lack a generally accepted definition. Recently, through extensive and thorough academic exchanges and exploration, the majority of tourism science workers prefer the following summarization: Tourism geography is a branch of human geography, and it studies tourism, recuperation and recreation of people and the geographical environments, and their relationships with social and economic development. We think this definition is the most scientific and specific description of tourism geography, which is the crystal of basic theoretical researches of tourism geography.

The relationship between tourism and geographical environment has been under study overseas for nearly 20 years, as proved by representative works of many scholars, such as *Water Resources and Tourism* by Tanner in 1973, *Vegetation Environment and Tourism* by Goodall in 1975, *Recreation and Tourism at Coastal Zones* by Glyptis in 1979, *Island Environment and Tourism* by Harrison in 1983, *Taking Maldives for Example to Discuss the Exploitation of Tourism Resources of Island Environment* by Manfred in 1985 and *Functions of Tourism on Land Use of Mediterranean Coastal Zone in Spain* by Tyrakowski et al. in 1986. Chinese scholars such as Guo Laixi, Yang Guanxiong, Chen Chuankang, et al.,

---

[2] Department of Tourism of Beijing International Studies University, 1983, lecture notes of *Tourism Geography*.

have expounded many times the relationship between tourism and geographical environment as well as social and economic development, which provides a universally recognized basis for the research on the concept of tourism geography.

1.2 Research Tasks of Tourism Earth-science

The tasks of tourism earth-science are to explore and study various earthscientific tourism resources by virtue of the basic theories, methods and technical means of earth-science; to discuss the distribution rules, texture and structure, formation mechanism and evolution of all sceneries; to verify, evaluate, plan and design tourist areas and spots and specific sceneries, in order to exploit, transform, use and protect the tourism resources; and, finally, to provide more, better, well-formed and rich earthscientific tourism resources for the tourism industry, and lay a firm and reliable material foundation for the development of tourism. Therefore, the fundamental task of tourism earth-science is to create and improve sightseeing places which are easily accessible, agreeable in landscape, informative and interesting. The specific research tasks are set out as follows:

1.2.1 Survey and Evaluation of Tourism Resources

Tourism resources are the bases where tourists can appreciate, spend holidays, recuperate, entertain, relax, venture into unknown and strange things and investigate. In other words, they are the object—one of the three elements of tourism—and are the material basis of tourism development. We should find out the status quo of tourism resources, including their types, quantities, sizes, grades, quality, characteristics, causes of formation, utilization, values and functions. This task is especially important in China. Only after making clear the conditions of resources can we work out a comprehensive and practical tourism development plan and make objective studies on the tourism development strategies and goals of a district. The task of tourism earth-science workers is to cooperate with planning, urban construction, and tourism administration authorities to check out the tourism resources in different areas, scientifically classify them, discuss their reasonable utilization channels and potentials, and make down-to-earthscientific and technical certification and evaluation.

1.2.2 Selection, Layout, Planning and Construction of Tourist Areas (Spots)

The selection of the tourist areas (spots), the size (capacity) of tourist reception, the arrangement of tourist accommodations, transportation and service facilities, the organization and supply of tourism goods, the layout of enterprises in tourist areas, and the balanced protection of tourism environments and ecosystem are all significant earth-science problems during the construction of tourist areas and should be studied in depth.

The selection and planning of tourist areas or spots fall into two basic types: the first type is new tourist spots, such as Changli Golden Coastal Resort in Hebei Province, tourist areas in Xingcheng of Liaoning Province, Yantai Yangma Island Travel Resort of Shandong Province, Wuxi Lu'er Mountain Scenic Area of Jiangsu Province and Stone Lake Scenic Area of Suzhou; and the second type is the rebuilding and expansion of the old tourist areas. Since the opening up and rebuilding of the tourist areas involve many economic departments and industries (including urban construction, transportation, ecosystem, environmental protection, catering, culture and entertainment, craft and special local products as well as living services), the selection and planning of the tourist areas must highlight priorities and achieve an overall balance in line with the regional planning of the whole area. Especially, the tourism development strategy of a tourist area must be integrated into the overall strategic development plan of the whole city or district. Of course, it depends on the specific conditions of the said district to determine whether it should be taken as the strategic objective, task or strategic priority. For example, there is a significant difference between Shanghai and Xi'an in the importance of tourism in their overall strategic plans.

1.2.3   Rules of Geographic Distribution and Spatial Movement of Travellers

The research on the occurrence and change rules of tourist markets involves a range of earthscientific issues, such as geographic factors leading to tourism, and the earthscientific backgrounds, especially the climate, beach, sunshine, view, landform, scenic water and scenic forest, which stimulate tourist flows. Besides, the geographical distribution and movement of tourist flows are often controlled by some special rules. Such rules govern the tourist movements from cold areas to warm areas, from humid and rainy areas to sunny areas, from developed areas to poor areas, from the city to the country or from the country to the city. Research is also made on the change and forecast of tourist markets, tourism seasons and characteristics of tourist activities in different areas. Tourist markets will change with any of the factors (especially the dominant factors). Some tourist markets are target markets, while some others are potential ones. Some target markets may recede as secondary markets, while some potential markets may rise as target markets. Therefore, the research on changes of the tourist markets provides important information for running tourism well. Tourism managers can adjust operation guidelines and contents at any time according to the changes in tourist flows.

1.2.4 Impacts of Tourism Development on Regional Economic Complexes

The development of tourism will promote all the economic sectors in a region. In connection with industry, it will promote the development of manufacturers of tourism-related food products, daily necessities, goods, arts, crafts, etc.; in connection with agriculture, tourism-related agricultural by-products and local specialties will benefit from tourism; in connection with commerce, tourism will activate circulation, advance commodity economy and speed up logistics and market exchange by linking up production and marketing. Moreover, tourism has a great influence on transportation, tertiary industries as well as employment. All these will result in a change in the layout, composition and proportion of the existing economic complexes. Such an influence is an important subject of tourism earth-science.

In addition to the aforesaid economic impacts, tourism will also change the geographical environment of a region. With the exploitation of tourism resources, all the facilities, hotels, merchandise and real estate serving tourism will gain development momentum, changing desolate places into densely populated communities; with the construction of tourist roads, a tourism-featured ordinary town may become a bustling area for entertaining tourists. In fact, these changes are ascribable to the modification and utilization of the geological and geographical environments, which are also an important subject of tourism earth-science.

1.2.5 Tourism Zoning

A tourist area is a geographical unit composed of several sightseeing places and scenic spots. Such a geographical unit is not only easily accessible geographically but also consistent in the combination of landscape types and touring routes. According to zoning principles, tourism earth-science workers divide a country or a region into many relatively independent but interrelated tourist areas, which is a useful guide for lateral contact and cooperation of tourist spots, route design and organization of tourist activities. Meanwhile, it is the primary basis for a country to make an overall strategic plan for exploiting tourism resources and developing tourism. Therefore, the research on the zoning of tourist areas is the basic work of tourism social system engineering and has its theoretical significance as well as practical value.

1.2.6 Tourist Transportation and Other Relevant Means

The position and function of tourist transportation in tourism development, and the characteristics, function evaluation and spatial layout of various means of transport are all important subjects of tourism earth-science. Since transportation is the tie and media of tourist flows, it is one of the basic elements of tourism and the direct link between man and scenery. Therefore, modern large-scale tourism could not be formed or developed without transportation.

Tourist map is a kind of themed functional map. It mainly serves tourist guide and tourism earth-science and is an indispensable tool for tourists. It has strengthened the connection between people and landscapes and made touring activities more convenient for tourists, and it is also a guide for choosing the best touring routes and searching the best landscapes as well as a consultant for self-guided tours. Functions, types and drawing method of tourist maps are one of the research tasks of tourism earth-science.

1.2.7 Tourism Resources and Environmental Protection

The development of tourism will bring "public benefits" to the society as well as "public nuisances" to the environment. Tourism earth-science workers should not only pay attention to the exploitation and utilization of tourism resources in order to develop its "public benefits", but also study the influence of tourism on the environment in a bid to protect the environment and ecology during tourism development and correctly and scientifically balance the protection of tourism resources and environments and the development of tourist areas. The research covers the types of "public nuisances" of tourism; the modes, channels and results of the influence of "public nuisances" on the environment; the ways and measures of protection of tourism environments; and evaluation and forecast of environmental quality and relevant technologies.

1.2.8 Basic Theories and Methods of Tourism Earth-science

Tourism earth-science, as a science studying the interrelation between people and landscapes, should have its own complete theoretical system and research methods. We should sum up those theories and methods in practice so as to enrich, improve and sublimate them.

In addition to the aforesaid research contents of tourism earth-science, tourism geology and tourism geography, as branches of tourism earth-science, have their respective unique research tasks. Tourism geology focuses on (1) research on the causes of formation of geology and landforms of landscapes and places of interests. The research contents cover time, space, materials and sports as well as formation mechanisms. It takes regional geological environment as the background and combines the geological and geomorphical conditions of an area to elaborate the formation process. It tries to make qualitative and quantitative observation and analysis of ground features studied, such as "a line-like sky", "flying stone", "sword stone" and "peach stone", whose positions, heights, directions, sizes, attitudes, lithology, geological conditions, causes of formation as well as evolution processes should be observed. (2) Geological research in the exploitation and utilization of scenery resources, such as geological survey of scenic areas, which can be made from the perspectives of point, line and plane. The survey from the perspective of point is limited to tourist spots with centralized landscapes, simple geological conditions and a small area; the survey from the perspective of line is made along touring routes; and the survey from the perspective of plane is adopted to draw geological and geomorphic maps, profile maps and sketch maps with landscape elements, and specimens and samples like rocks, paleontological fossils, soil and water will be collected for indoor identification and assay if necessary. Moreover, corresponding geological work is also done for geologically relevant museums, repositories and parks. (3) Geological research on protecting tourism resources and tourism buildings. It is the unique research task of tourism geology to protect the stability of scenery resources, in such respects as preventing stone scenery from gravity-collapse; preventing cave scenery from corrosion and weathering; maintaining the water temperature, water quality and gushing water quantity of springs; and preventing carved stones of a cliff from exfoliation and corrosion. This is of great, special research significance for protecting scenery resources and guaranteeing the safety of tourists.

1.3 Historical Development of Research on Tourism Earth-science

Tourism earth-science is an emerging discipline. It is born from the marriage of earth-science and tourism science when modern tourism develops to a certain phase. Tourism earth-science overseas dates back to the 1930s at the earliest, while in China, it emerged in the late 1970s to the early 1980s. Tourism earth-science is a growing discipline. The term "tourism earth-science" was first put forward by Chinese scholars in 1985. Its first definition was included in Article 2 of "Articles of China Tourism earth-science Research Association", "Tourism earth-science is an emerging marginal discipline which uses the theories and methods of earth-science to serve the investigation, research, planning, development, transformation and protection of tourism resources". The term or complete concept of "tourism earth-science" has not yet appeared overseas. In that sense, Chinese scholars are the founders of tourism earth-science as a complete discipline. In Sect. 1.1, we have proposed the new concept of tourism earth-science: Tourism earth-science is a generic term for two marginal disciplines—tourism geology and tourism geography. The term "tourism geography" was put forward by foreign scholars in the 1930s, while China started systematic research on tourism geology and tourism geography relatively later.

1.3.1 Research History of Tourism Earth-science Abroad

Although the term "tourism earth-science" has not been generally accepted internationally, related researches began quite early in the world. Tourism geography researches overseas began in the 1920s to the 1930s, which is a sign and product of geographers entering modern tourism research field. Modern tourism originated after the 1840s. Although its formation aroused attention of many scholars, it is generally examined by social economic scholars from the perspective of economic development. It is just 60–70 years since geographers began to study modern tourism from a geographical perspective, and *Relationship between Recreational Activities and Land Use* by K.C. Mcmurry was generally recognized as the first work of geographers studying modern tourism. However, owing to incorrect understanding of the concepts "tourism" and "entertainment" in the academic circles, theoretical research on tourism geography has been at a standstill for a long time. American scholar P. Schnell held that "tourism" may be understood in a broad or narrow sense, with the broad sense being "included in a generic term of various recreational activities"; I.M. Matley puts "tourism" under "recreation", while R. Fussell treats the two words as synonyms. J.B Jackson argues that "Tourism is mainly a geographical experience and is a free and relaxed way in which we learn about the world, ourselves and our lifestyles". Such confused understandings of tourism have directly led to the difficulty in agreeing upon the research scope of tourism geography. Although British scholar Brown pointed out as early as in 1935 that geographers should pay more attention to tourism research, yet for quite a long time, most works on tourism geography focused on some tourist resorts or the significance of tourism economy but seldom discussed the basic theories of tourism geography.

In the 1950s, some geographers integrated tourism geography into economic geography and mainly studied tourism from the perspective of economic attributes because tourism is an economic undertaking. In 1964, Roy. L. Wolfe, a Canadian geographer, held that tourism geography should be taken as an independent science since it was separated from economic geography, and he had his *Sightseeing Places of Ontario* published in 1960. (Laixi 1985) C.A. Stansfield had his *Seashore Summer Resorts of America published.* (Mingde 1988) H. Robinson, a British geographer, took tourism geography as an applied geography. I.M. Matley, an American geographer, pointed out that "European geographers pushed up tourism to the level of erudition". However, Japanese geographer Asaka Yukio still included tourism geography into economic geography in his *Tourism Geography*, while some other Japanese geographers took it as a branch of cultural geography. Although the above understandings of tourism geography disagree widely, it shows that people has been attaching great importance to the disciplinary attributes and theories of tourism geography.

After a series of argumentation on the disciplinary attributes of tourism geography, a number of works on tourism geography came out in the world. Some of the influential works are as follows: the thesis titled "Taking Ito as an Example to

Discuss the Composition of Population and Labor Movement of Tourism Cities"
by Japanese scholar Shiro Suzuki Fu in 1958; the article titled "Relationship
Between Recuperation and Entertainment Forms of Residents and Other Areas"
by Japanese Yoichi Koike, published in *Geographical Review* in 1960; the article titled "Recuperation and Entertainment Land, the Land for the Future" by
M. Clawson of the USA in 1960, in which M. Clawson classified the recuperation
and entertainment land into three types: type of user tendency, resources-based
type and intermediate type; the article titled "Analysis of the Formation Factors
of Tourism Industry—A Case Study of Guanping Ski Resort" (Zhang 1985) by
Qingyi Rongyi and Yiteng Daxiong of Japan in 1962; *Geography and Tourism
Research* by Italian geographer Nice Brouno in 1965, in which he proposed five
research subjects of tourism geography: basic impetus of tourism, tourism environment and space, the influence of tourism on residence of people, application
of tourist maps, and tourism planning (Nice 1965); in the same year, UK scholar
Stansfield published an article titled "Complementary Method of Division of
Resort Entertainment Functions" in *Tourist Review* in 1965, arguing that there are
three standards for the division of tourism cities: (1) ratio of tourism employees
to the all the employees of the city; (2) the per capita entertainment income in
the city; (3) the proportion of personal entertainment income to per capita income
(Stansfield 1965); and the thesis titled "Influence of Tourism Facilities on Tourism
Hinterlands" by G.F. Deasy and R.G. Phyllis of the USA, in which they pointed
out the differential attraction scope of tourist towns with identical geographical
bases is ascribable to the external distance of tourism markets and internal facilities and publicity (Zhang 1985).

In the late 1960s to the early 1970s, essays published by scholars around the
world also include the following: "Research into Tourism Geography" by Yantian
Xiaosan of Japan (1968), "Geographic Location of Tourism Movement" by
K. Ruppert and J. Maler of Germany (1970), "Tourism Localization of Coastal
Fishing Village—A Case Study of Kii Long Island Town" by Shancun Shunci and
Hashikura of Japan (1971), "Water Resources and Tourism" by Tanner in 1973,
and "Vegetation Environment and Tourism" by Goodall.

In 1974, Japanese geographer Asaka Yukio came out with his monograph on
tourism geography named *Sightseeing Geography* (Asaka 1980), which had significant influence around the world.

H. Robinson had his *Tourism Geography* (Robinson 1976) published in 1976,
which focuses on the description of tourism development, evolution of human
needs, tourism motivations and factors, measurement and scope of tourism, tourism and environment, tourism organization, transportation, economic and social
significance, tourism planning as well as regional tourism development around the
world.

Jcleno Ambroxio, a geographer of Yugoslavia, had his *Tourism Geography* published in 1979, in which he pointed out that the tasks of tourism geography were
to study origins of tourism, possibilities and conditions of human tourism, attractive
bases of tourism, touring routes and tourism types, facilities of different tourist spots,
confirmation of potential and promising tourism areas, laws of tourist flows, etc.

E.A. Kotpob, a Soviet geographer, had his *Recreation and Tourism Geography* published in 1978, which mainly elaborates the theoretical bases of formation of regional recreation complexes, the preconditions for formation and development of regional recreation complexes, zoning and evaluation of recreational land, distribution of regional recreation complexes, natural protection and transformation of recreational land.

The Geological Institute of the Academy of Science of the USSR together with seven higher education institutions compiled *Geography of Recreational Systems of Soviet Union* in 1980, which elaborated the conditions of recreational activities and tourism resources and functions of recreational geographical domain systems of Soviet Union, and geographical domain organizations, and divided the Soviet Union into 4 recreational zones and 20 recreational areas.

G.H. Pirie, a geographer of South Africa, wrote *Tourism Data and Time-Space Ecology* (Pirie 1979) in 1979, in which he proposed research on various tourism data from the ecological perspective of time and space in order to avoid environmental pollution arising from tourism.

American geographer Britton wrote a book named *Geography of Leisure* (a new concept), in which he included international tourism, domestic tourism, recreational tourism, urban recreation as well as sports, making the controversial differences between "tourism" and "recreation" limited to the difference in function only. Therefore, the research domain of tourism geography was further extended.

I.M. Matley wrote *The Geography of International Tourism* in 1976, a representative one of the numerous works on international tourism geography since the 1970s.

Major works on tourism geography published in the 1980s include the following: *Marine Tourism Geography and Utilization of Marine Tourism Resources* (a chapter of "Marine Economic Geography") by E.T Zhasilafuski of Soviet Union; *Review on Current Tourism Recreational Places* by Shancun Shunci of Japan; *New Tourism Space of Civic Centres* by Pu Tatsuo of Japan; *Tourism Geography—Overview and Prospects* by L.S. Mitchell; *Changes of Recreational and Tourism Space around the Cities* by Shiro Suzuki Fu of Japan; and *General Development Direction of Tourist Areas and Zones of Soviet Union* by Pu Luobu Reski of Soviet Union.

In recent years, *Recreation Geography: Research on Location and Tourism* by Canadian famous scholars, L.J. Stephen and L.J. Smith, had great influence on the academic and theoretical researches on tourism geography around the world. Now, this book has been planned to be translated and published in China.

To meet the development requirements of tourism geography, international geographical organizations have held a succession of geographical meetings since the 1970s and made a series of decisions on the position of tourism geography. At a professional meeting on tourism geography convened in Austria in 1973, the International Geographic Society discussed the terminology, nature and development of tourism, tourism as a development factor of a country or region, the relationship between national boundaries and tourism development, and spatial behaviours of recreational activities. Tourism geography was firstly included into the professional group at the 23rd International Geographic Congress held in Moscow in 1976. This grouping principle was continued at the 24th International Geographic Congress held in Tokyo of Japan in 1981. Since then, tourism geography has been established as a branch of geography.

1.3.2 China's Research History of Tourism Earth-science

China's thoughts on tourism earth-science traced far back into history. Some ancient travellers, literators and poets covered simple earthscientific knowledge in travelogues, essays and scientific poems, such as *Mountain and Sea Classics*, *Buddhist Records*, *Travel Notes of Xu Xiake and Introduction to the Scenery of Beijing*, which can be regarded as China's budding thoughts on tourism earth-science. The Opium War forced open China's door, through which modern western sciences including earth-science entered China. The modern western earth-science took root in China, developed into an influential basic discipline and branched out into geology, geography, meteorology, etc. However, before the late 1970s, as the tourism industry in China started late, the earthscientific circle did not attach importance to tourism researches. Scholars only wrote some articles about earth-scientific knowledge for scenic areas but seldom made any in-depth research. This period witnessed the growth of China's modern earth-science and laid a theoretical foundation for the birth of tourism earth-science in China. The party's opening up and economic stimulus policies brought tremendous changes to the political and economic situations in China. The rapidly growing national economy, increasing international contacts and flooding foreign tourists have sparked China's tourism industry like a prairie fire. The status of tourism in the national economy has been rising day by day. Tourism has been drawing widespread attention and arousing great interest among scholars of earth-science, so that a number of geology or geography workers participated, voluntarily or upon invitation, in the research on some earthscientific issues involved in the tourism industry. By then, earth-science workers began to really set foot in the tourism industry, combining earth-science with tourism. The period from 1978 to 1985 when China Tourism earth-science Research Association was founded can be considered as the germination stage of tourism earth-science in China. The period from 1986 to 1991 is the start-up phase of tourism earth-science, and *The Introduction to Tourism earth-science* is a phased summary of China's research findings of tourism earth-science of the stage. The period from 1992 to 2000 was the growing stage for tourism earth-science. The theory and methods of tourism earth-science have become the motive force of pushing ahead the consideration of geoparks in china that was an import symbol in the period. The period from 2001 to present is a new development stage of the tourism earth-science. With the increasing in number of geoparks, the social demands for tourism geosciences are increasing also and the theory and methods of tourism earth-science are maturing. The publication of *A Grand Tourism earth-science* compiled mainly by Prof. Chen Anze (editor in chief) is an important symbol of the stage. Now we will sum up the brief history of the China Tourism Geoscience as follows:

1. *The Germination Phase of Tourism Earth-science in China (1978–1985)*
   China's national economy started to recover after 1976, injecting vitality and vibrancy throughout the country, including magnificent landscapes and waterscapes which have been dormant for many years. The tourism industry gathered unprecedented momentum to the shocking surprise of other industries. Of course, this shock also rolled on to geological and geographical circles, who felt obliged to do something for tourism. Against such a background, the academic community under years' ban shook off its shackles, and the Geological Society of China also resumed activities in 1978. To spread geological knowledge to the masses, the Geological Society of China set up a science popularization organization (with a preparatory group established in 1978, and Science Popularization Committee of Geological Society of China officially established in 1980), which conducted large-scale science popularization activities. Youth Summer Camp to Study earth-science, a good form for spreading geological knowledge to adolescents, came into being. The camp was tried out in Beijing in 1978 and had by 1981 spread to 29 provinces, municipalities directly under the central government and autonomous regions. Every year, thousands of youngsters make geological explorations at dozens of field camps. In a few years, the camps added up to several hundred, most of which are located in natural scenic areas. To make campers get systematic scientific knowledge, the organizers of the camp have compiled scientific guidance materials for the activity area of each camp. The guidance materials mainly include brief descriptions and overviews of the basic conditions and features of the geology, landform, cultural geography and economic geography of the camp. They are not only well received by campers but also take the fancy of common tourists. On this basis, Chen Anze and Li Weixin compiled *Exploring Secrets of the Earth*, a collection of short theses of middle school students, which has received good response. We can see that geology has infiltrated into tourism, and many geology workers have begun to address tourism issues. In 1979, *Geological Series of China's Places of Interest* with Professors Yin Weihan as editor in chief was published, and by 1985, the series had included science popularization and guidance materials focusing on geology of scenic areas such as Nanjing, Hangzhou, Jinan, Jiuhua Mountain, Mount Huangshan, Heaven Lake of Mount Changbai, Chengde, Xi'an, Shanghai, Beijing and Mount Hua, having some impact on earthscientific circle. The above works generally aim to spread geological knowledge and fall far short of development need of tourism. To discuss how the geological science can serve tourism better, Science Popularization Committee of Geological Society of China have held several small forums in Beijing, Xinjiang and Hunan since 1980. Tourism geological issues were discussed at the inaugurating meeting of the 2nd Science Popularization Committee of Geological Society of China in 1984. The attendees brought their research findings relating to tourism services in several years, including

the investigation results of tourism resources of Manjiatan in Dalian, Liaoning and Zhangjiajie Sandstone Peak Forest in Dayong, Hunan, which drew attention from attendees. Since then, the tourism service initiative of the earthscientific circle has entered the phase of investigating new tourism resources. The geographical circle has compiled a range of tourism geography textbooks and has planned and evaluated several tourist spots and areas. By then, the time had become ripe for China's earthscientific circle to hold a scientific symposium on tourism issues. Therefore, in the spring of 1985, the first symposium on tourism earth-science of China was held in Beijing as initiated by the Science Popularization Committee of Geological Society of China. The term "tourism earth-science" was first proposed by the planners of the symposium. As soon as the term was proposed, it was accepted by the geological and geographic circles as well as closely linked circles such as gardening, architecture, environmental protection, tourism, archaeology and museums. This academic conference was very successful and received 82 essays, which cover discussions on basic theories of earth-science, investigation reports on tourism earthscientific resources, and suggestions on tourist areas of different levels and maps of tourism resources. It was the first grand meeting to review China's research findings of tourism earth-science. The symposium provided an opportunity of mutual exchanges and leanings for the geological and geographic circles and the related academic circles. "Preparatory Committee of China Tourism earth-science Research Association" was set up at the symposium in order to continue such interdisciplinary exchanges and unite tourism researchers of the whole earthscientific circle. Sun Daguang, the then Minister of Geology and Mineral Resources, Chairman Huang Jiqing of the Geological Society of China, and famous geographer Hou Renzhi were elected as honorary chairmen; Chairman Chen Anze of Science Popularization Committee of Geological Society of China was elected as chairman; and Li Weixin was elected as secretary general. The association was subordinate to the Geological Museum of China. China Tourism earth-science Research Association, whose founding is a milestone in the research history of tourism earth-science of China, has provided an organizational guarantee for the establishment of the discipline of tourism earth-science in China, laid a solid foundation for the development of tourism earth-science, and opened a new stage of China's tourism earth-science.

2. *The Establishment Phase of China's Tourism Earth-science (1986–1991)*
The founding of the association has created favourable conditions for the research on tourism earth-science. As an academic organization committed to establishing a marginal discipline, it has since inception been calling upon its members to do more practical work, directly serve the development needs of China's tourism industry, sum up experience in practice and sublimate it into theories, and make every effort to create a tourism earth-science with Chinese characteristics. To that end, the association undertook it as its first task after establishment to draw a national tourism resources map for State Planning Commission and Bureau of Land and Resources. Thanks to the efforts of the

members throughout the country, it took only three months to complete drawing China's tourism resources map (1:4,000,000), which cover over 3000 tourist spots (areas) developed or to be developed and 717 important tourist sites, and prepared a detailed resource registration form. At the same time, a prospective zoning map of China's tourism resources (1:6,000,000) was also drawn. The association proposed to develop 50 tourist areas, with detailed instructions attached. The association also gave suggestions on guidelines of development of China's tourism resources, "With regard to tourism objects, the policy to be adopted should emphasize both domestic and overseas tourists and put particular emphasis during specific implementation according to specific circumstances; in terms of development and planning of tourist areas, we should focus on developed coastal areas, build priority sites in the inland, control the development of tourist areas in the west and shift the focus to the western region gradually during the "Ninth Five-Year Plan" period; as for the types of to-be-developed tourism resources, we should emphasize both cultural resources and natural resources, but should focus on cultural resources in the early stage, and then shift the focus gradually to the development of natural resources in the medium to long term; in respect of the general plan of the exploitation of tourism resources, we should have the idea of building integrated tourist areas by breaking the constraint of administrative districts, centring on a major city or important scenic areas and in combination with traffic conditions, and avoid building isolated tourist spots so as to give full play to the economic benefits of tourism facilities; in the near term, we should focus on coastal regions and the east and build a batch of seashore tourist resorts in coastal regions; and, in the inland, we should set up a batch of integrated tourist areas with unique characteristics according to resource characteristics, based on natural traffic conditions and relying on central cities. We should also pay attention to building a batch of national nature parks. In western regions, distinctive tourism resources should be exploited selectively, in a bid to meet foreign tourists' need of seeking novelty, explorations and expeditions, but extensive exploitation should be conducted later. The central government and local governments should give play to their initiatives in exploring new tourism areas. Before official development by the central government, the local governments should be allowed to conduct primary development and utilization in a planned and directed manner, but should not rush into mass action to prevent damaging the united layout and tourism resources. Under the guidance of the overall strategic plan of the state, a long-term plan should be prepared in order to gradually form a comprehensive system of tourist areas throughout China. The research findings, which provided reference for State Development Planning Commission to prepare a long-term plan for national land rectification, won high praise. To meet the development need of the discipline, the association proposed the plan to compile *The Principles of Tourism Earth-science*. At the 2nd annual academic conference of tourism earth-science held in Wuhan in the winter of 1986, issues relating to drawing of tourism earthscientific maps were reviewed, and the compilation rules and graphic designs of maps of tourism resources in Zhejiang, Guizhou, Inner Mongolia, etc., were given high credit. Aerial photographs of tourist cities and scenic areas displayed at the conference unveiled a new world for attendants. Several outlines of *The Principles of Tourism Earth-science* were discussed in detail, compilers were determined, and members were mobilized to review the outlines and write articles for the 3rd annual academic conference. At the 3rd annual conference of tourism earth-science convened as scheduled at Mount Tianmu, Zhejiang, in the autumn of 1987, 67 theses were received, which focused on the basic theories of tourism earth-science, including research objects and tasks of tourism earth-science, classification and evaluation of tourism resources, aesthetic evaluation of tourism earthscientific resources, research on regional development of tourism resources, examples of investigation and evaluation of tourism resources. These theses marked a series of research findings relating to the basic theories of tourism earth-science. The excellent theses received at the past three annual academic conferences were included in *Tourism Earth-science Album.* The 4th annual academic conference convened at Xi'an in the autumn of 1988 was themed on the investigation and evaluation of tourism earthscientific resources, with 80 theses received. Compared with those received at previous conferences, these theses reflected more profound research findings and proposed a set of procedures and methods of the investigation of tourism earthscientific resources, made qualitative description and quantitative analysis in evaluating resources, and put forward the scheme of making evaluations by mathematical models, laying a theoretical foundation for the evaluation of tourism earthscientific resources. In the recent decade, the research on tourism earth-science has been centring on serving tourism, and a lot of related practical work has been done. During the period, a variety of highly valuable tourism resources such as Zhangjiajie, Nine-village Valley and Wild Three Hills were discovered and developed, and a lot of tourist areas have also been evaluated and planned. A preliminary theoretical system of tourism earth-science has taken shape. Meanwhile, a research team of tourism earth-science has been set up as well; members of the association added up to more than 500 and are distributed all over China; and six provincial tourism earth-science research associations were set up in Xinjiang, Sichuan, Fujian, Hubei, Zhejiang and Shaanxi, with more under preparation, showing a scene of prosperity of China's research on tourism earth-science. All provincial associations frequently organized activities and came out with a range of high-level theses and works, as represented by *Research on Tourism Earth-science and Development of Tourism Resources* of Sichuan Provincial Tourism Earth-science Research Association, as well as *China's Tourism Earth-science Series* with Cui Guangzhen and Chen Anze as editors in chief. The geographical circle has made a lot of researches on tourism issues and has made a series of research findings and completed many works. Many tourism institutions and departments, such as China Tourism Institute, Beijing Tourism Institute, Tourism Department of Nankai University, Tourism Department of Xibei University, Department of Tourism Economics of Hangzhou University and Shanghai Institute of Tourism, have been endeavouring to study tourism geography and tourism earth-science, and tourism geography or tourism geology has been listed as a required course or an elective course in many tourism institutions and the department of geography or geology of multiversities and normal colleges. The academic circle and publishing circle have compiled and published a series of monographs and textbooks in respect of tourism earth-science, such as *Chinese Tourism Geography* by Zhou Jinbu of Hangzhou University, *Tourism Geography* with Liu Zhenli of Beijing Tourism Institute as editor in chief, *Tourism Geography* with Lei Mingde of Xibei University as editor in chief, and *Modern Tourism Geography* compiled by Lu Yunting of Beijing Normal University. It is noteworthy that Professor Lin Chao and Professor Chen Chuankang of Peking University have been tutoring China's first masters of tourism geography from 1982. By 1986, seven masters of tourism geography had graduated in China, becoming the first generation of professional tourism earth-science workers. At present, they are working for the scientific research and education of tourism earth-science and have compiled a series of high-level research theses on tourism geography.

The first and foremost thing to do after the founding of the preparatory association was to set about establishing the discipline of tourism earth-science; for there had been no ready discipline of tourism earth-science before the term tourism earth-science was proposed, and if an academic body was not supported by a mature discipline, this academic body would be very difficult to keep on carrying out academic activities for long. It is, therefore, imperative to progressively establish an entirely brand new discipline of tourism earth-science with an independent research object, a complete set of working methods and a solid theoretical foundation. The association determined that starting with practice, it would go along the road of "practice, theoretical research, again practice, and improving the theory on tourism earth-science" and meanwhile called on all its members to take an active part in the practical activities of serving the tourist industry and conduct academic discussion centring on the establishment of the discipline of tourism earth-science. Through great efforts of all its members, plenty of practical data, work methods and practical work experience were accumulated; thus, the time to write the monograph *An Introduction to Tourism Earth-science* became ripe gradually. Under Prof. Chen Anze's auspices, the first monograph of tourism earth-science titled *An Introduction to Tourism Earth-science* was published by the Peking University Press in 1991. In a preface to this monograph, Huang Jiqing (then President of the Geological Society of China) wrote "*An Introduction to Tourism earth-science*, compiled by Chen Anze, Lu Yunting et al. of China Tourism earth-science Research Association, is a summary of the research achievements of several hundred members of the Association in the five years since its establishment, as well as a pioneering move of China's earth-science workers to serve the tourist industry and apply the theory and methods of earth-science to the tourist industry". As soon as the book *An Introduction to Tourism Earth-science* came out, it was chosen by many universities and colleges as teaching material or a must-read reference book, which played an important role in training talented people of tourism and guiding the development of the tourist industry in terms of both theory and practice in the initial founding period of tourism earth-science. The publication of *An Introduction to Tourism Earth-science* marked initial founding of the discipline of tourism earth-science.

3. *Growing Period of Tourism Earth-science (1992–2000)*

In this period, academic activities concerning tourism earth-science in China became increasingly frequent; the results of its services to tourism were more and more conspicuous; its influence on the tourism community of the whole country was progressively expanded; and it commenced to forge ahead onto the international stage. During this period, a series of important events took place: (a) beginning from 1992, the China Tourism Earth-science Research Association (Preparatory) formally became a 2nd-level organization of the Geological Society of China, China Tourism Association and Domestic Tourism Association, and, when the Association communicated with foreign colleagues, the word "preparatory" was deleted and the name China Tourism Earth-science Research Association was adopted. (b) The concept of the tourism geological (earth-science) industry was put forth for the first time. On 12 March 1992, Song Jian, then State Councillor of the State Council, wrote a letter to the then Ministry of Geology and Mineral Resources (MGMR) concerning how to develop the geological tourist industry, and the MGMR asked the China Tourism Earth-science Research Association to discuss implementation measures. Headed by Chen Anze, the association undertook a soft science project of the State Science and Technology Commission entitled the "Research on the Developmental Strategy of the Tourism Geological (earth-science) Industry in China", which comprehensively and systematically dealt with problems involved in the development of the tourism geological (earth-science) industry in China. This implies that the work of tourism earth-science had attracted the attention of state leaders. (c) The discipline of tourism earth-science began to step into Taiwan and onto the stages of international geological conferences. On the occasion of celebrating the 70th anniversary of the establishment of the Geological Society of China in 1992, Ruan Weizhou, a geologist of Taiwan, received the book *An Introduction to Tourism Earth-science*, which was brought back to Taiwan and given as a present to Prof. Wang Xin of the Department of Geography, Taiwan University. This means that tourism earth-science was then introduced into Taiwan. Prof. Wang invited Chen Anze to Taiwan University in 1995 and 1997, respectively, to give a lecture about tourism earth-science, which made it possible for tourism earth-science to be disseminated in Taiwan. In 1996, when the 30th International Geological Congress was held in Beijing, Chen Anze and Chen Maoxun delivered a thesis entitled *Tourism Earth-science—A New Field of Earth-science*, in which the connotation of tourism earth-science and its development in China were for the first time introduced to over 6000 earth scientists from 120 countries or so attending the congress. Quite a number of members of the association also plunged themselves into selection of more than 80 routes of geological field trips of this congress and their related organizing work. Altogether, over 1000 foreign geologists participated in the field trips, which made tourism earth-science tested at the highest scientific level. (d) The association organized and held the symposium on the Strategy of Development of Tourism Resources in the western region of China. With sensitive perception, the association predicted that the era of large-scale development was going to arrive in the western region. As nearly as 1994, the association had gathered experts in tourism earth-science to discuss problems concerning the strategy of developing tourism resources in the western region, analysed systematically the characteristics of tourism resources and present state of their development in the western region and put forward the strategic objective and measures for the development of tourism resources in the western region, which made positive contributions to the development of the tourist industry in the western region of China. (e) A large number of young earth-science experts emerged. The tourism earth-science community attached much importance to fostering young experts in tourism earth-science, and during this period of time, the first group of masters and doctors with tourism earth scientists as directors graduated from the Chengdu College of Geology.

4. *New Development Period for Tourism Earth-science (2001–)*
When the dawn of the new epoch shed its light on the land of China, we ushered in a brand new development period of tourism earth-science. The emergence of China national geoparks, which were established at the suggestion of tourism earth-science experts, is the most prominent mark of this period. They open up a new service field for the discipline. In this period, the emphasis of work and research concerning tourism earth-science has begun to expand to services to the construction of geoparks, with symposia on geoparks held, a collection entitled *Construction of Geoparks and Development of Tourism Resources* published, many geopark research projects carried out and a series of international academic conferences on geoparks held. The theory and methods of tourism earth-science have become the theoretical basis for guiding the construction of geoparks, and the contingents of tourism earth-science have become the core force in the construction of geoparks. Through research into tourism earth-science, plenty of new data were accumulated, which laid a good foundation for the revision of *An Introduction to Tourism Earth-science* and compilation of *A Grand Tourism earth-science Dictionary*. In order to link tourism earth-science activities with construction of geoparks more closely, the Geological Society of China made a decision that the China Tourism earth-science Research Association shall expand and be renamed the China Association of Tourism Earth-science and Geopark Research attached to the Geological Society of China; in addition, a non-governmental research institution called the Chinese Academy of Tourism Earth-science was also set up in 2010 to strengthen research on tourism earth-science. At present, tourism earth-science has been included by many tourism colleges or geological universities and colleges as an elective or compulsory course and quite a number of postgraduate students of tourism earth-science have been turned out. Tourism earth-science and tourism earth-science workers have made great contributions to the development of the tourist industry of China and have been highly praised by the China National Tourism Administration. The theory and methods of tourism earth-science have become the motive force of pushing ahead the construction of

geoparks in China, and furthermore, they have also been adopted and pursued by departments in charge of land and resources. With the publication of the English edition of *An Introduction of Tourism Earth-science* and completion in compiling *A Grand Tourism Earth-science Dictionary*, the discipline of tourism earth-science has preliminarily stepped onto the international academic stage. The appearance of geoparks is an important mark of this period, and by inheriting the past and ushering in the future, tourism earth-science is forging ahead into a mature period.

Although tourism earth-science has been recognized as a discipline both at China and abroad, it is still quite young and expected very much to be improved in terms of theoretical research, content and methodology, and particularly, it needs to be disseminated and extended both in China and in the world so as to attract the attention and concern of more people and receive their support. Therefore, the emergence of geoparks and arrival of the scientific era of tourism in China give tourism earth-science an opportunity to climb a new high peak and predict a bright future.

References

Asaka, Y. (1980). *Tourism geography*. Tokyo: Damingtang.

Cui, G. (1989). Review on China's research on tourism earth-science. *Discovery of Nature*, 8(2), 44–45.

КоТπЯроВ, Е. А. (1978). ГеорафиЯ ОТπыIX и Туи3а, MockRa.

Laixi, G. (1985). *An emerging branch of human geography—Tourism geography, Collected Essays of Human Geography* (p. 284). Beijing: People's Education Press.

Lu, Y. (1988). *Modern tourism geography* (pp. 2–3). Nanjing: Jiangsu People's Publishing House.

Matley, I. M. (1976). *The geography of international tourism*. Washington: Association of American Geographer.

Mingde, L. (1988). *Tourism geography* (p. 17). Xi'an: Xibei University Press.

Nice, B. (1965). Geografia e Turism Ricerca, Geografica Ituliana, Vol., X X II, No. 3.

Pirie, G. H. (1979). Travel data and spatiotemporal ecology. *The South African Geographical Journal, 61*(2), 119–122.

Robinson, H. (1976). *A geography of tourism*. Macdoald & Evans, London.

Shunci, S., & Hashikura, (1971). Tourism localization of coastal fishing village—a case study of Kii Long Island town. *Collected Essays of Dato Bunka Great Economy, 17*, 53–77.

Stansfeld, C. J. (1965). Complementary method of division of resort entertainment functions, *Tourist Review*, 148–157.

Xiaosan, Y. (1968). *Research into tourist geography*. Tokyo: Ming Xuan Study.

Zhang, W. (1985). *Bases of economic geography* (p. 201). Ji'nan: Shandong Science and Technology Press.

Chapter 2
Role of Tourism Earth-science in Tourism Development

2.1 Tourism Earth-science and Tourism System

2.1.1 Tourism System

Tourism is a huge social systematic project comprising tourism demand and supply systems and many subsystems, such as tourism regional system, tourism market system, tourism service system, tourism goods system and tourism education system. These parts, which are interdependent and infiltrate and constrain each other, constitute a complicated tourism architecture, as presented below:

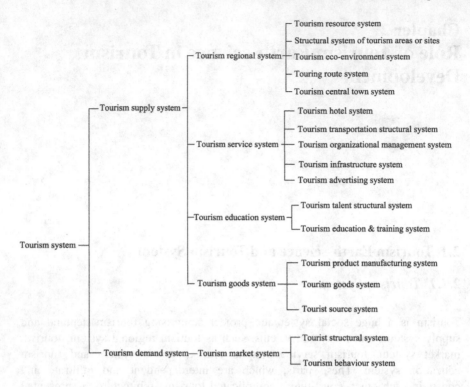

By functions of tourism, the aforesaid systems can be merged into three aspects: subject, object and medium. The subject is the tourist, who is the leading actor of a tourism activity and without whom tourism will not happen; the object is the tourist site or scenic resource, which is the material basis for attracting tourists and without which tourists will not be stimulated or lured and tourist activities will not happen; the medium, also called bond, is the intermediate link between the subject and the object and serves to enhance the accessibility of tourism by means of service, so it is also indispensable to tourist activities. The aforesaid three aspects are the basic elements of tourism. They are interrelated, constrain each other and form an organic tourism complex. Tourism economics, tourism marketing, tourism management, tourism earth-science, tourism psychology, tourism literature, tourism aesthetics, tourism law, etc. under tourism science study this complex from different perspectives and objects. Therefore, tourism earth-science is not only an important component of tourism science, but also a branch of earth-science. It mainly studies some things and phenomena relating to earth-science, namely undertakes the earthscientific task of studying tourism complex and its systems.

2.1.2 Earthscientific Background of Tourism System

Tourism occurs in a certain spatial setting. All elements constituting tourism are inevitably merged into this spatial system, forming a tourism regional complex. Therefore, tourism system is pregnant with rich earthscientific characteristics. Also, because region (including the entire earth surface) is the basic stage of earthscientific research activities, only such an entity may help reveal the laws of many earthscientific phenomena that happen in this region. Tourism earth-science is a discipline which studies the relationship between tourism and geological and geographical environments. Therefore, all tourism events and phenomena in the region certainly have rich and profound earthscientific contents. Now, we will analyse the earthscientific characteristics of all elements constituting a tourism system.

2.1.2.1 Earthscientific Characteristics of the Subject of Tourism

The subject of tourism—tourist is the source of tourism demand and an important condition to enhance the recognition of tourism sites. Tourists start from their starting points, travel along specific touring routes by different means of transport to destination countries or destinations and then return to the starting points or go to other tourist areas, which constitute the whole process of the tourism market system. Each link has research contents of earth-science. For example, the excitation conditions of tourism are inseparable from the different natural and cultural environments of a tourist area. Those conditions may trigger in tourists' various tourism motivations linked to politics, economy, culture, history, aesthetics and psychology. The environmental conditions on the earth surface including climate, beaches, sunshine, topography, geology, water areas, living things can all induce tourists. Some visitors motivated by the desire for knowledge intend to study the scientific value of these conditions and to expand their visions of knowledge; some motivated by the desire for pleasure intend to appreciate natural beauty and express feelings by virtue of sceneries; some want physical exercise and mental relaxation in clean and quiet natural environment and fresh air; some seek variations from their living routines to meet their psychological need for adventure; and some hope to avoid bad weather in their hometowns and seek warmth and sunshine. We can see that all these different tourism motivations are closely related to certain environmental features of the earth surface. Tourism earth-science workers will certainly cover a lot of research contents of earth-science while they are studying these tourism causes and motivations.

The regional differences of tourist sources not only affect the travel motives of travellers but also determine the scale structures and geographic movement patterns of tourist flows. Research subjects of tourism earth-science are the trends of tourist flows from cold areas to mild ones, from wet rainy areas to bright sunny ones, from hot plains to mountains and coastal summer resorts, from developed areas to developing zones, from the countryside to the city and from the city to the countryside. In a word, the difference of tourism regional conditions is a basic factor for the imbalanced time–space distribution and movement intensity of tourist flows and is an important condition for the alternation of high season, shoulder season and low season of tourism in various areas. Tourism earth-science workers have been diving into the characteristics of tourism seasons and tourist activities in various areas and the difference in use value of the same tourist site in different seasons, which is a compulsory research task of tourism earth-science.

The research and prediction of changes in tourist source markets also bear some relation to tourism earth-science. It is known to all that the distribution of tourist sources and the direction of tourist flows are affected by many environmental factors, with some of their changes directly determined by earthscientific environments. Therefore, if a certain factor attracting tourists changes, tourist source markets will show new trends and characteristics. Tourism earth-science studies such subtle market changes and is therefore very informative for tourism managers in making decisions to adjust economic policies and contents.

To sum up, the tourism market system includes many subsystems such as tourist source system, tourist structure system and tourism behaviour system, all of which contain rich research contents of tourism earth-science. Tourism earth-science studies these earthscientific factors and characteristics that restrain tourist source markets, which is undoubtedly of great significance for the development of tourism.

2.1.2.2 Earthscientific Background of the Object of Tourism

The object of tourism—tourism resources, which include many subsystems such as resource classification system, tourist site structure system, tourism eco-environment system, touring route system and tourism infrastructure system—is the basic material condition for the formation of tourism regional system through attracting travellers. These systems not only share common earthscientific background but also have unique structural characteristics, and therefore, they are the main study object of tourism earth-science and are of huge practical value for the overall development and economic growth of tourism regional system.

Tourism regional system is also known as tourism recreation system abroad. It is the general term for regional series with tourism functions. The basic characteristics of tourism regional system are as follows: (1) Spatiality: namely the system has corresponding spatial scope and definite boundary of attraction; (2) substantiveness: namely the system is a complex of a series of regions with tourism objects and tourism facilities; (3) hierarchy: namely the system contains several levels of recreational regions which interrelate and constrain each other, with high-level recreational regions restricting low-level recreational regions and low-level recreational regions forming part of high-level recreational regions;

(4) domination: namely there exists the difference in dominant tourism functions between each level of the tourism regional system and its adjacent levels. In other words, each tourism region in the tourism regional system has its own dominant functions and advantages, which makes it become a unique tourist site competing with its surrounding scenic areas. In a word, tourism regional system is a series complex of multifunctional, multi-level and multi-category tourist zones, areas, lines and spots with systematic characteristics. The said series complex mainly includes the following research contents: location factors for the formation and development of tourism regional system; unique functions of tourism regions; resource characteristics of tourism regions; economic development level and environmental conditions in tourism regions; distribution pattern of travel centre in all recreation regions; and facilities and organizations in tourism regions.

Tourism resource classification series is the basic element constituting tourism regional system. Tourism resources are a general term for natural and cultural resources which are attractive to tourists and have certain tourism functions and values. In other words, they are absorbing materials with certain tourism functions and values. It can be seen that tourism resources, regardless of natural landscapes such as mountains, water, climate, trees and flowers or cultural landscapes such as city walls, palaces, temples, gardens, tombs, ancient towers and modern architectures, are all scenic entities and phenomena distributed in certain regions. Their natures, types, sizes, grades, environments and factors are closely related to earth-science, and some things and phenomena are the basic research objects and contents of earth-science.

Structural series of tourist sites is an important part of the object of tourism. According to the functional system, economic structure and formal role of tourism supply and demand, this series is divided into many types such as landscape tourist sites, cultural tourist sites, historical tourist sites, ethnic tourist sites, modern engineering tourist sites, entertainment and recreation tourist sites and comprehensive tourist sites. Landscape tourist sites can be further divided into sightseeing tourist sites, summer tourist sites, winter tourist sites, spa tourist sites and sports tourist site. These different tourist sites have the following earthscientific characteristics: (1) They belong to a region of certain location and scope; (2) there are specific and exploited tourism resources distributed in the said region for sightseeing; (3) the economic structure in the region is dominated by tourism, namely there are comprehensive tourism supply facilities and services. All these are the research contents of regional earth-science.

Tourism ecological series is also part of earth-science worth study in the object of tourism. Ecological series is a general term for all elements which constitute eco-environment. How many ecological elements are involved in a tourist area depends on its environment location. The ecological elements of city parks mainly include water, soil, atmosphere and biology; the ecological elements of nature reserves mainly include topography, water, soil, atmosphere and biology. A change in any of these ecological elements will have a certain impact on the whole eco-environment. Therefore, the beauty of environment is the product of the organic integration of various ecological elements. In order to protect the beauty and pleasure of tourism environment, it is necessary to learn the rules of ecological changes in tourism environments from the perspective of earth-science so as to artificially maintain ecological balance and promote a virtuous circle of tourism environments, and give full play to the comprehensive effects of various factors such as natural resources background, secondary resources background, social and economic conditions and science and technology.

Touring route series is the artificial design engineering bound up with the elements of tourism earth-science and plays a critical role in the development of tourist areas or tourist sites. It is known to all that tourism is a kind of regional spatial movement of tourists from the departure place to the destination. This temporary movement to other places is restricted by various factors such as tourist areas (spots), economic status, traffic conditions, tourism markets and vacation time. However, all spatial movements for the purpose of sightseeing, recuperation and recreation require the selection of a certain route. Such a route should be economical, comfortable and multifunctional, featuring versatile and distinctive sightseeing contents; convenient accessibility for tourists to avoid roundabout and return; touring spots appropriately spaced to make tourist activities properly intervaled and rhythmic; consideration of hot and cold spots to maintain the balance of tourist flows. It can be seen that the selection, planning and development of touring routes are also the major limiting factors and conditions of tourist activities. Tourism earth-science workers should put their professional theories and methods into practice to design more travel routes which are convenient, efficient, fast, safe, comfortable and economical.

In a word, the above series or systems are the basic research contents of the object of tourism. Their formation, structure, layout and range of action are all based on the theories of tourism earth-science. Comprehensive study on the object of tourism is indispensable to the development of tourism.

2.1.2.3 Earthscientific Characteristics of the Media of Tourism

The media of tourism, also known as the tie of tourism, includes a series of intermediate links such as tourism hotels, restaurants, transportation facilities, communication conditions, travel agencies, travel companies, tourism translation, guidance, publicity, insurances and laws. Its basic functions are reception, management and service. Different from the above subject and object, these items of media are all the artificial design systems serving tourists. The research on them is generally included in the domain of cultural research, but many aspects of them have to appeal to earth-science, or in other words, some research contents of them need to extend to the research field of earth-science. For example, with regard to site selection for hotels and restaurants in tourist sites, a series of geological and geographical factors like geographical location, geological foundation, environmental engineering and natural aesthetics are involved in many earthscientific issues and need to be studied. Geographical location is important; in that, it is related to the customer source effect of hotels and restaurants; geological foundation should be solid in that it is related to the lifetime, safety and utilization of buildings; environmental engineering is the basic condition for hotels and restaurants because it is related to the investment scale, construction pace and impacts on surroundings of these buildings; natural aesthetics is an important indicator of environmental protection and the capacity, form and colour of buildings must be in harmony with the natural eco-environment.

Tourist transportation also includes many research subjects related to earth-science. The selection of type of tourist transportation, the construction conditions of transportation facilities (parking lot, cableway, etc.) as well as the requirements for the roadbed and road surface of tourist roads all need certification by professionals engaged in earth-science. Because the determination of transportation types such as railways, roads, helicopters and cableways entails scientific research on the corresponding location, geology, topography and environment. Otherwise, blind construction will lead to unexpected problems such as waste of investment, inadequate tourist sources, environmental destruction and hidden appendix-type risks. Site selection for parking lots and cableways should also be based on a review on earthscientific issues such as geological foundations, geomorphic conditions and environmental impacts. Moreover, roadbeds and road surfaces as well as directions and materials of walkways are directly related to environmental conditions such as lithology, structure, gradient and vegetation. It can be seen that the layout and building of transportation facilities need to be evaluated and certified by tourism earth-science workers.

And tourism publicity, guidance, translation and various drawings are also closely related to earth-science, because geology, geomorphy, water, climate and biology are not only the research objects of earth-science but also the landscapes directly appreciated by travellers. In order to make tourists rapidly and profoundly obtain the information of sceneries and enhance the attractiveness and recognition of them, it is necessary to disseminate the information through publicity, guidance and translation. It can be seen that the research contents of tourism earth-science are just the knowledge that should be acquired during these intermediate links.

2.2 Effects of Tourism Earth-science in Tourism Development

Like other tourism sciences, tourism earth-science promotes the development of tourism in various aspects. A few decades ago, some countries with developed modern tourism had noticed the pivotal position of tourism earth-science in the development of tourism and had conducted a special study on the scope and means of tourism earth-science so as to give full play to the tourism guidance function of this discipline. In 1939, Professor E.W. Gibbert, a British geographer, came up with a theory of "the functional relationship between geographical factors and tourism growth", and formulate this relation as follows:

$$F = f(x)$$

In the formula, x represents a variety of earthscientific factors: Geology (x_1), geography (x_2), resource (x_3), land use (x_4), climate (x_5), transportation (x_6)... It can be seen that the development of earthscientific factors is positively proportional to tourism growth. Years of experience further show the following effects of earth-science on tourism development.

2.2.1 Survey and Evaluation of Tourism Resources

Tourism, as a special lifestyle of residents, basically features leaving one's own residences for other places for temporary sightseeing, recuperation, holidaying and recreation so as to seek beauty, pleasure, knowledge and novelty as well as go shopping. Therefore, tourism resources are the basic material conditions for travellers' sightseeing and adventure. In order to find out the background and visual, scientific and historical values of tourism resources, it is necessary to survey and evaluate the distribution, characteristics, quantity, quality, origin and exploitation conditions of earthscientific tourism resources. As earthscientific tourism resources fall within the domain of earth-science and have earthscientific characteristics, only professionals acquainted with earth-science may penetrate into their nature and explore tourism functions and values through using earthscientific theories and methods. It can be seen that the survey and evaluation of regional landscape resources are one of the compulsory research tasks of geologists and geographers as well as the strategic work to provide material bases for tourism development. Through survey, observation and research, tourism earth-science workers can not only understand the characteristics, type, value, geology, geomorphy, climate, hydrology, vegetation, transportation and economic conditions of tourism resources, but also write the survey results into systematic reports and prepare various drawings by professional means. In recent years, many Chinese earth-science workers have written very featured reports and prepared sets of drawings about tourism resources, such as distribution maps of scenic spots and areas, classification and zoning maps of landscape resources, research and development maps of landscape resources and zoning and forecast maps of landscape resources. These textual and graphic results comprehensively reflect the general characteristics of tourism resources in a region or a scenic area and provide reliable scientific bases for overall regional development, planning and design.

The comprehensive evaluation of tourism resources is the central part of the research on tourism earth-science. This work is based on a general survey of the resources and a comprehensive evaluation report is basically required to be made on the abundance, quality, grade, composition characteristics, visual value, cultural value, scientific value, environmental quality, exploitation conditions and comprehensive benefits of tourism resources through an analysis of adequate survey data. There are various types of evaluation, such as qualitative evaluation, quantitative evaluation, integrated evaluation from both qualitative and quantitative perspectives, demand evaluation, supply evaluation and the integrated evaluation with demand and supply combined. These evaluations are not only the basis for preparing the exploitation plan of tourism resources but also the primary means to improve the recognition of scenic spots. Such work is the incumbent research task of tourism earth-science.

2.2.2 Discovery and Selection of Scenic Development Zones

The development of tourism requires scientific combination and support of many scenic areas of different types, characteristics and functions, namely comprehensive scenic areas, as well as special scenic areas with major functions of sightseeing, scientific exploration, sports, recuperation, summering, display of historical cultures and ethnic customs, religious and ancestral pilgrimage. These scenic areas (spots) with different exploitation functions and values are needed by travellers of different countries, ages, sexes, occupations, educational levels, interests and preferences and are important conditions for improving the attractiveness and exploitation rate of tourism. Tourism earth-science workers ought to be good at discovering all kinds of scenic areas so as to serialize tourism development zones in a country or a region. Discovery of new scenic areas and spots is not only a prerequisite for the development of tourism resources but also an important task for further developing the new and old scenic areas already opened to the public. It is known to all that tourism is a business with heated competition. If a moderately developed old scenic area gets stuck in a rut and pays no attention to subsequent development, its tourist sources will be grabbed by tourist areas with richer resources, greater attraction and competitiveness. Therefore, all old and new scenic spots already opened to the public need to keep developing themselves by finding new and beautiful tourism resources. In recent years, many new scenic areas have been discovered nationwide such as Nine-village Valley, Huanglong Temple and Xingwen Stone Forest Cave in Sichuan; Manghe River and Mount Yawu in Henan; Tenglong Cave, White Horse Cave and Mount Gaolan in Hubei; Laoling, Wild Three Hills and Changli Golden Coast in Hebei; Yangma Island and Liangshan Cave in Shandong. These newly discovered scenic areas may support the tourism layout of various regions. Some scenic spots were isolated in the past, but a good touring route formed with newly discovered local and surrounding scenic sites have turned sluggish cold spots into profitable hot spots. Moreover, with the discovery of new attractions, some old scenic spots have given themselves a facelift by including more sightseeing contents, thereby attracting swarms of travellers from around the world. Mount Emei, a Buddhism shrine in Sichuan, is still deficient in supporting facilities of landscape resources in spite of the existing extensive development. Afterwards, earthscientific scenic spots such as caves, structures, mineral springs and waterfalls were discovered through further exploration, which supported the intensive development of Mount Emei.

There are various ways to discover new tourism resources and select new scenic sites by using earthscientific knowledge and methods. For example, some new scenic spots can be explored and discovered by virtue of the clues provided by local chronicles, past travel notes, heritage information, public reporting of landscapes, extension rules of formation conditions of the existing scenic areas, zoning forecast report of tourism resources, as well as modern scientific means including satellite photography and aerial remote sensing. The unique contribution of tourism earth-science is undoubtedly ascribable to the perfect presentation of its strengths.

2.2.3 Precise and Rough Measurements of the Area and Quantity of Elements of Natural and Cultural Landscapes in Scenic Areas

A scenic area or sightseeing city is made up of different types of landscapes, geological and geographical elements. According to our survey data in recent years, there are at least 100 earthscientific landscape elements. Basic landscape elements include land surfaces, water areas, highlands, plains, plateaus, hills, basins, deserts, grasslands, woodlands and shoaly lands; geological and geomorphic elements of landscapes include stratigraphic sections, typical structures, fossils, earthquake relics, special mines, ancient mining sites, volcanoes, peak forests, caves and glaciers; meteorological and climatological elements of landscapes include sunlight, sky colours, heat, rain, fog, mirage and Buddha light; aqueous elements of landscapes include rivers, lakes, sea, springs, waterfalls, pools and rivulets; botanical and zoological elements of landscapes include forests, old trees, precious trees, flowers, green fields and ornamental animals; cultural elements of landscapes include cities, towns, countryside, factories, bridges, ports, as well as palaces, halls, buildings, pavilions, houses, alcoves, platforms, porches, pavilions on a terrace, monasteries, temples, tombs, pagodas, gardens, stone carvings, inscriptions, steles, couplets on pillars, murals, grottos, historical monuments and former dwellings of celebrities. These elements have different measurement requirements: Some require precision and some require estimation; some require a regular quantity and some require special measurement. Take plant species for example. In addition to the number of families, genera and species, it is better to list the statistics of dominant species, endemic species, rare species, protected species and ornamental species too. Such statistics are of great significance for understanding the tourist features and functions of some scenic area or nature reserve and are the fundamental basis for designing tourism facilities and the measures for protection of the ecosystem. For example, Mount Huangshan has

72 peaks, 24 springs, 20 pools, 16 rivulets, 14 caves, 3 waterfalls, 2 lakes, more than 2400 species of plants, nearly 100 species of fish and amphibians, more than 200 species of birds and animals, 14 species of animals under state protection; Maiji Scenic Area of Tianshui has 4 main landscapes, 18 small scenic areas, 24 independent scenic spots, 7 ancient sites, 194 grottos, more than 7200 statues and over 1000 m^2 of murals. In a word, such landscape statistics are the results of earthscientific researches by predecessors and are fundamental to a comprehensive understanding and evaluation of a scenic area. Earth-science as a comprehensive discipline can not only make scientific measurements of various earthscientific objects and phenomena through surveys but also make scientific "diagnoses" to identify the authenticity of the statistics so as to make the measurement of scenic elements more accurate and systematic.

2.2.4 Positioning, Qualitative and Quantitative Analyses of Intrinsic Attributes of Tourism Resources

Apart from external features and categories, some intrinsic attributes of landscape elements are the internal determinatives of landscapes and are essential for the development of sceneries. A scenic area not only gives sensuous enjoyment to tourists, but also enriches their knowledge. Only if external beauty is integrated with internal essence can we deeply understand the visual, scientific and cultural values of sceneries, thereby sublimating the appreciation and experience of such a beauty. For example, the external elements of karst scenic areas are just peak forests, stone columns, stone cones, stone buds, stone caves, solution caves, funnels, canyons, natural bridges and waterfalls. However, upon careful comparison, we may find that the karst scenic areas have different characteristics in the nature of sights, scale, shape and development stage. From the macroscopic perspective, their geological foundations are karst, but why are there differences in shape? It is necessary to do some positioning, qualitative and quantitative analyses on the changes in the combinations of karst components in various scenic areas.

Positioning refers to the environmental location of each type of landscape in the scenic area. This location is the result of the integrated action of various geographical factors of nature and is organically related to its surrounding environmental factors. For example, Beijing Cave of Stone Flower, which happens to be located at the south foot of Dashi River, has developed better than Cloud and Water Cave of Mount Shangfang in the south, and it is impossible to find such a solution cave in more southerly places like Juma River Basin and Shidu Scenic Spot. All these are related to the environmental location, indicating that the paleoenvironment of the Cave of Stone Flower was somewhat different from that of Cloud and Water Cave and Shidu Scenic Spot.

Qualitative analysis means the identification of the essential properties of landscapes. Every landscape has its specific scientific name. For example, peak forest usually refers to a forest made up of limestones and its essential characteristic is dissolution process of carbonate rocks by acidic water containing carbon dioxide. However, there are some false peak forests which are formed by endogenic and exogenic geological processes, instead of the dissolution process, on other sedimentary rocks or intrusive rocks. It can be seen that different peak forests have various development factors. Peak forest is a unified form of landscape with different natures and properties. Earth-science workers have been endeavouring to make a scientific "diagnosis" of the essential attributes of peak forests since they have found that peak forests are similar in appearance but different in essence. For another example, cave landscapes are uniformly named as grotto from the perspective of appearance, but they vary greatly in the type of internal sights, spatial structure, scale and causes of formation. According to their formation mechanisms and lithology, they are generally classified as limestone cave, volcanic cave, granite cave and rhyolite cave. Solution cave and fluxing holes have specific meanings. The former is a karst-specific phenomenon, its lithological compositions must be carbonate rocks, and its formation mechanism is mainly corrosion as well as auxiliary gravitational collapse; while the latter is a phenomenon unique to volcanic lava, its lithological compositions must be extrusive rocks (like basalt) and the formation mechanism is mainly imbalanced condensation of lava and specific auxiliary conditions like underlying landform. It can be seen that qualitative research on caves must be based on special earth-scientific knowledge. Particularly, limestone caves have quite complicated types of landscapes. In respect of nature, there is cave geology, cave sediments, cave water, underground rivers, inside waterfalls, springs, hot mineral water, cave life, fossils, and outside geomorphy, hydrology, vegetation, etc.; in respect of culture, there are ancient buildings, stone inscriptions, stone carvings, wall writings, cave cultural sites, myths and legends. Therefore, survey and evaluation of caves is specialized work involving many research contents of earth-science such as exploring the trend and scale of solution caves, measuring the shapes, finding out the types, quantity and quality of shapes and studying the development conditions and value.

Quantitative analysis refers to the quantitative relationship of landscape components. Every landscape is inseparable from the interactive influences of solid, liquid and gas. Cave landscapes are mainly formed by solid rocks under the interaction of liquid water and air. Mechanical composition, contents of chemical elements, stratigraphic bedding, the trend and size of joint and fault lineaments of rocks all have a certain impact on the formation, scale and shape of solution caves. These items can be quantitatively analysed with modern scientific experimental methods.

2.2.5 Supervision and Protection of Tourism Resources and Their Ecological Environments

Environmental issue is one of the major strategic issues of the world today. The protection of human living environment has aroused widespread attention around the world. The current massive environmental pollutions not only affect the survival of human beings, endanger animals and plants and lead to depletion of natural resources, but also pose a tremendous threat to tourism. Recent researches indicate that the development of tourism and environmental protection has been contradictory and interdependent in objectives. Beautiful environments are prerequisite to attracting travellers, so substantial development of tourism relies on landscape resources and their eco-environment. However, owing to the inherent characteristics of tourism, its development will more or less harm the healthy social and ecological environments and threaten the foundation that tourism itself depends on. It can be seen that the development of tourism can rapidly bring economic benefits from the perspective of single economic unit but it is detrimental to environmental protection; and conversely, the damage and spoiling of environments directly affect the sustainable development of tourism in the area. The contradictory and interdependent relationship between tourism development and environmental protection determines the duality of economic strategy research of tourism. That is, tourism benefits should be obtained through rational development, and protection of resources and environments should be strengthened so as to maintain stable and sustainable development of tourism. As mentioned earlier, tourism earth-science is a science which studies the relationship between human touring activities and geological and geographical environments, so it specially focuses on the protection and purification of tourism environments. In order to protect tourism resources and environments from destruction, tourism earth-science can study and monitor tourism environments from different perspectives. For example, how does tourism bring bad impacts to specific environments? How to determine the ecological capacity limit of different landscapes? What impacts will the destruction of social and ecological environments bring to tourism? What policies measures should be enacted to protect tourism resources and eco-environment? In a word, tourism earth scientists should attach great importance to the research on these issues and use modern scientific and technical means to strengthen supervision and forecast of environmental changes in scenic areas on the basis of evaluation of environmental quality of scenic areas.

2.2.6 Preparation of Tourism Development Plans

Tourism plan is a crisscross matrix plan with multiple levels, multiple programs and disciplines. The comprehensive coordination and action of various disciplines are the basic guarantee for successful planning. As tourism earth-science is an emerging science which studies the rules of regional landscapes by virtue of theories and methods of earth-science, it plays an irreplaceable role in the practical process of tourism planning. Particularly, a series of issues in regional tourism planning have to be addressed under the guidance of theories and methods of earth-science. For example, the discovery and evaluation of tourism resources, the capacity measurement of scenic resources and environments, the exertion and combination of functions of limited regional resources, the scientific interpretation of the causes of formation of tourism resources, the rational organization and design of touring routes and the scientific layout of various sections in tourism complexes, all have distinct earthscientific connotations and need to be studied and solved by earth-science workers. It can be seen that tourism planning is the specific application of earthscientific theories and methods in the domain of earth-science as well as an important practice of the research and development of tourism earth-science.

The contents and methods of tourism planning will be studied in detail in Chap. 8.

Chapter 3
Basic Formation Conditions of Natural Tourism Resources

3.1 Significance of Research on Formation Conditions of Natural Tourism Resources

3.1.1 Research Contents of Formation Conditions

Natural tourism resources are natural landscapes and natural environments available for enjoyment by man. They reside in a certain spatial position, specific formation conditions and historical evolution stages of nature. The research contents of their formation conditions include the following:

1. The formation and location of landscapes in the spheres of the earth, the interaction of respective spheres and the relationship with the integrated natural geographical environments such as regional geology, landform, climate, hydrology, soil and living things.
2. The geological age, the position of earth structure and palaeogeographic environments of the formation of landscapes in the historical evolution process of the earth, as well as conditions like the strata, rocks, geological structures and geological dynamic process for the formation of landscapes.
3. Characteristics of land hydrology, ocean dynamics and ocean geography of the formation of landscapes.
4. Solar radiation and atmospheric circulation conditions of landscape formation and emergence, as well as the relationship between landscape formation and emergence and meteorological factors, and climatological characteristics as well as ground structures.
5. The relationship between landscapes and regional biological evolution, biogeography and human activities.

© Springer-Verlag Berlin Heidelberg and Science Press Ltd. 2015
A. Chen et al., *The Principles of Geotourism*, Springer Geography,
DOI 10.1007/978-3-662-46697-1_3

3.1.2 Significance of Research on Formation Conditions

The survey on the distribution and formation conditions and emergence rules of the known natural tourism resources and landscapes is aimed to (Yunting 1988; Writing group of tourism introduction 1983):

1. Determine the formation modes, evolution and distribution rules of natural tourism resources and landscapes; draw the formation condition and distribution rule maps of national and regional natural tourism resources and landscapes; draw prospective forecast maps of natural tourism resources; and draw development technical conditions and prospective construction planning maps.
2. Study the formation conditions, history, evolution rules and development trends of natural tourism landscapes in each tourist area, in order to directly serve the construction and development needs of tourist areas.
3. Make clear the causes of formation and features of causes of formation of natural tourism landscapes and make scientific identification and evaluation, which will be used as the contents of scientific guidance and the channels to disseminate scientific knowledge, as well as the basic materials for deeper researches of professional scientific tourists.

3.2 Basic Formation Conditions of Natural Tourism Resources

3.2.1 Spheres of the Earth and Natural Tourism Resources

The earth on which man lives is one of the nine planets of the solar system and of course a member of the big universe. Earth surface can be divided into crust, biosphere, hydrosphere and atmosphere. Man is a member of the biosphere. Thanks to the development of science and technology, man can reach out into the space and down into the inside of the earth, climb up mountains or stay deep under the sea, and go on expeditions and explorations, with their domains of activities expanding continuously. Also, the natural landscapes for tourism and developable natural tourism resources have been constantly increasing. Various natural tourism resources can be formed in every sphere which human activities may reach, such as a variety of wonderful and deep underground geological tourism resources formed in the lithosphere and various geomorphic tourism resources formed on the lithospheric surface; shimmering water tourism resources like rivers, lakes and sea formed in the hydrosphere; dazzling and diversified biological tourism resources like animals, plants and micro-organisms as well as terrestrial and aquatic organisms in the biosphere; and changing meteorological tourism resources and climatological tourism resources like warm, humid, cool and hot weather in the atmosphere.

3.2.2 Regional Integrated Natural Geographical Environments (Kalesnik 1947)

Integrated natural geographical environments reflect the structures, causes of formation, dynamics and development rules of all natural elements on the entire earth surface. These natural elements include geology, landform, climate, hydrology, soil, vegetation and animals. Natural geographical factors correlate to and restrain each other and combine organically to form an internally consistent whole. The formation and evolution of certain natural tourism resources depend on a certain integrated natural geographical environment. For example, the four wonders of Mount Huangshan—fantastic pines, grotesque rocks, the sea of clouds and hot springs— are determined by the granite mountain landform with vertically developed joints, warm and humid climate, lush vegetation and specific conditions of spring effusion. On the contrary, without warm and humid climate, there would be no fantastic pines all over the mountains, and without granites with highly developed joints and corresponding landforms, there would be no grotesque rocks and hot springs. Therefore, during the research and evaluation of natural tourism resources, emphasis should be placed on the integrity of natural geographical environment elements.

Major integrated natural geographical regions determine the regional distribution rules of natural tourism resources. For example, in China's seven major natural geographical regions, different natural tourism resources have different natural landscape characteristics: (1) North-east China: with ice and snow, forests and volcanic landforms as major tourism landscapes; (2) North China: with tourism landscapes dominated by lofty and rugged mountains and sandy beaches; (3) Central China: with picturesque hills and lakes and subtropical vegetation as major tourism landscapes; (4) South China: with tourism landscapes dominated by tropical, subtropical integrated natural landscapes and picturesque Karst landforms; (5) Inner Mongolia–Xinjiang region: with grasslands, icebergs, deserts and Yardang landforms as major natural landscapes; (6) Qinghai–Tibet region: with mountain glaciers, plateau lakes and hot and bubbling springs as major tourism landscapes; and (7) Kangdian region: with mountains or canyons, rapid waterfalls, intermontane basins, downfaulted lakes and karst stone forests as major tourism landscapes.

3.2.3 Geological Structures and Crustal Structures (Geography Department of Nanjing University 1963)

The characteristics of crustal structure and geological structure of an area result from the crustal movement and evolution of geological environments of the said area, which determine the strata, rocks, structural features, magmatic activities, metamorphism and geomorphic forms in the area. The abovementioned factors to some extent determine the landscape types and formation of natural tourism resources.

From the perspective of global tectonics and crustal structure, the crust can be divided into oceanic crust, continental crust and transitional crust by type. The global crust is divided into six major plates and more than 20 small plates by active tectonic zones; different parts of the respective plates form various natural tourism resources under different dynamic geological processes. The volcano/earthquake-active island-arc zone consisting of the Aleutian Islands—Japanese Archipelago—Ryukyu Islands—Philippine Archipelago, formed by the subduction zones between Pacific Plate and Eurasian Plate, is a tourist area with natural landscapes predominated by ocean, islands, volcanos and hot springs. The collision of Indian Plate and Eurasian Plate results in the Qinghai–Tibet Plateau with the thickest crust, which is known as the third pole of the earth and is a natural tourist area with mountain glaciers as major natural landscapes.

From the perspective of tectonic units of the continental crust, different geotectonic evolution histories and geological environments form distinctive natural tourism resources. Continental shelf region with stable crust and even strata forms tourism geomorphic landscapes such as loess plateaus, sandstone peak forests and karst peak forests; geosynclinal region with intense orogeny forms natural tourism landscapes such as mountain glaciers and canyon turbulent flows; geodepressional region developed from the activated continental shelf forms extremely thick continental strata and complex magmatic rocks and natural landscapes like Danxia landforms and fault block mountains.

The characteristics of regional geological structures control the formation of natural tourist areas. Mountainous tourist areas such as magnificent Mount Taishan, Mount Hengshan, Mount Huashan and Mount Lushan are formed in the fault block uplifted areas, and downfaulted lake tourist areas such as Poyang Lake, Qinghai Lake, Dead Sea and Lake Baikal are formed in the downfaulted area; specific structural forms control the formation of scenic spots, such as the Dragon Head Cliff on Mount Lushan, which is a fault cliff, and the West Sea of Mount Huangshan, which is a set of granites intensively cut by vertical joints, and the Infatuation Stone on Mount Taishan, which is the diabase vein with developed annular joints.

3.2.4 Strata and Rocks (Zhengzhou Geology University 1979)

In the evolution history of the earth, extremely thick layered rocks and variously shaped rock bodies are formed, including layered sedimentary rocks, metamorphic rocks, volcanic rocks, magmatic rocks and mixed rock bodies. They record the whole history of the earth since its formation including palaeogeography, paleoclimate, palaeontological evolution, tectonic movement, magma and volcanic activities, mineralization, metamorphism and the impacts of cosmic events on the earth, which constitute a masterpiece of geological history.

1. Strata and palaeontological fossils in strata are all important tourism resources. Included in a national nature reserve, the Middle–Late Proterozoic stratigraphic

section in Jixian County with its nearly 10,000 m-thick strata records the 600–1700 million years' geological history of North China, which has important scientific tourism value and has been attracting a large number of travellers from around the world. There are also the state-level Shanwang strata and the palaeontological nature reserve in Linqu County, Shandong Province.

2. Because of the disparities in chemical compositions, physical properties, diagenetic features, structures, construction, minerals and combinations, different strata and rocks form different tourism geomorphic landscapes. Under certain conditions, particular tourism geomorphic landscapes will appear in specific strata and rock distribution areas. Only in large carbonatite strata distribution areas can we can find karst tourist attractions such as Guilin scenery, Guizhou Dragon Palace, Lunan Stone Forest and Fangshan Cloud and Water Cave; only in stratigraphical distribution areas of glutenites with flat attitude and developed joints can we find Danxia landform landscapes with fantastic peaks and grotesque stones; and only in solid granite distribution areas will there be perilous and miraculous tourism geomorphic landscapes such as Mount Huangshan and Mount Huashan.

3. Some rocks themselves may also form important tourism resources, such as Yuhua stones (rain flower stones) of Nanjing, calcareous tufa of Dongwushi Village, Cizhou, and the Helan stone, one of the five treasures of Ningxia. Furthermore, some worshippers who go to the top of Mount Taishan for religious tourism regard Mount Taishan stone as a kind of treasure, and they carry one piece back and put it at the head of the bed to ward off evil influence. When building a house in rural areas, people embed a Mount Taishan stone tablet into the gable to subdue evils. This practice of collecting Mount Taishan stone has been a motive of worshippers in the west of Shandong visiting Mount Taishan.

3.2.5 *Geological Dynamic Process*

Every natural tourism landscape has a natural history of formation, namely the evolution history of dynamic geological process, and is a result of the interaction between endogenic force and exogenic force of geological process.

Endogenic geological process is a force that originates inside the earth and can change the form of the earth surface and the characteristics of rocks, such as the upwelling and eruption of magma caused by volcanic process, crustal arching and downwarping as well as faults resulting from tectonic movement. A considerable number of natural tourism resources are formed under the endogenic geological process, such as the crater and volcanic landscapes of Heaven Lake of Mount Changbai, and the Northern Mount Heng formed by block uplifting.

Exogenic geological process is the force that originates in the atmosphere, hydrosphere and biosphere outside the crust and can change the form of the earth surface and the characteristics of rocks, such as weathering, gravitational collapse, erosion, transportation and accumulation. A good many natural tourism landscapes are formed under exogenic geological process, like stone forests, solution caves

and peak forests among Karst landforms, as well as the Yardang landform—Ghost
City, formed by wind erosion.

However, most natural tourism landscapes are formed under the geological
interaction between endogenic and exogenic forces, such as the natural landscapes
of Three Gorges which are formed by the latest arched uplifting and continuous
erosion and down-cutting of rivers, and mountainous tourism resources of magnificent and rugged Five Famous Mountains which are formed by block uplifting
and discrepant erosion.

3.2.6 Land Hydrological Characteristics and Ocean

The surface of the earth is surrounded by a continuous hydrosphere. The total
volume of the hydrosphere is 326 billion km^3, 97.2 % of which is distributed in
the oceans. On the land surface, the hydrosphere appears in the forms of rivers,
lakes, glaciers, springs and streams. It penetrates and converts mutually with the
atmosphere and lithosphere, moves rapidly and varies endlessly, forming a series
of tourism wonders in nature and providing varied comfortable environments and
rich imageries for human beings.

Glaciers, rivers, lakes and springs on the continent's surface form distinctive
land hydrological landscapes and rich natural tourism resources on seven continents in the world under the impact of geology, geomorphy and climate. Most of
the major rivers in Asia originate in central alpine areas, radiate out in all directions and flow into the marginal seas of the Pacific Ocean, the Indian Ocean, the
Arctic Ocean and the Atlantic Ocean, respectively, forming natural landscapes
such as mountains and valleys, hills and plains, and lakes and swamps, as well as
mountain glaciers and plateau lakes of inland waters in continental interiors.

In Europe, the land and ocean are indented, and the river network is dense, with
stable flow and long navigable mileage, forming natural tourism landscapes such
as the blue Danube and the clear and bright Rhine.

In western America, the Cordillera is near the Pacific Ocean so that the ocean-
bound rivers originating in the Cordillera are very short, while in the open east,
there appear the world famous falls tourist areas such as the Niagara Falls and the
Angel Falls in the Great Lakes region. Rivers of Africa have a series of huge waterfalls, like 72 waterfalls on Zambezi River, resulting from different hardness of strata
in the river courses, complex geological structures and numerous terraces and cliffs.

Oceans, totalling 362 million km^2 in area, account for 71 % of the earth surface. Generally, the central part is called ocean, the part adjacent to the land is
called sea, and the part stretching into the continental interior is called inland sea.
The formation of marine tourism resources is subject to ocean-type crustal and
geological structures, submarine landform characteristics, marine climate and currents, and marine lives. Beach tourism resources are formed at the edge of land,
while islands, coral reefs and marine lives, as well as marine wonders like sea fire
and sea light, are formed in the ocean.

3.2.7 Latitudinal Zonation and Regional Factors of Climate

Climate is the comprehensive characteristics of weather in a region for years, which are determined by interactions of solar radiation, atmospheric circulation and the properties of underlying surfaces. It is also a determinant factor of exogenic geological process. It has a controlling effect on the sculpture of landforms, formation of scenic waters and the growth and evolution of ornamental creatures in natural tourism resources. In particular space and time, environments with agreeable weather are themselves important natural tourism resource, such as summer resorts on mountains and beaches and tropical and subtropical winter resorts. Under certain climatic conditions, varied physical and chemical phenomena in the atmosphere including cold, heat, dryness and moistness, as well as wind, cloud, rain, snow, frost, fog, thunder, lightning and light, are all natural tourism resources, such as ice and snow in Harbin, the sea of clouds on Mount Huangshan, fog and cloud on Mount Lu, Buddha's light on Mount Emei, mirage in Penglai and singing sand in Dunhuang.

Differences in the climate on the earth surface are primarily subject to solar radiation. Climatic characteristics of a region are mainly related to its latitude. Different latitudes bear different intensities of solar radiation, leading to latitudinal zonations of global climate, where varied natural tourism resources such as tropical rain forests, ice and snow in frigid areas are formed. Secondly, the characteristics of atmospheric circulation, which are subject to land and sea distribution on the earth surface, large landforms and the intensity of regional solar radiation, have great impact on the transmission and exchange of atmospheric heat, energy and vapour, thereby influencing the regional variation of global climate. Therefore, despite the same latitude, great climatic variations and various landscapes can be formed due to the differences in distance from ocean, in coastline shapes and in geomorphic structures. For example, in the north temperate zone, the offshore areas feature tourism climatic landscapes of broad-leaved forests, while the inland regions far from the ocean are dominated by tourism climatic landscapes of deserts. Thirdly, vertical climatic zonations formed by partial mountainous landforms, that is there are different tourism climatic landscapes at different altitudes. Temperature, its diurnal range and annual range decrease progressively with the increase in altitude. For example, Xincun, Tangdan and Luoxue in Dongchuan City, Yunnan, with the relative altitude of 1000 m record annual average temperatures of 20, 13 and 7 °C, respectively, thereby forming different natural landscapes. Owing to the vertical variations of mountain climate, there are many mountain summer resorts in East China, such as Mount Lu, Mount Jigong and Mount Mogan, which record an average annual temperature range of 5–8 °C with the plains below the mountains.

3.2.8 Biogeographical Characteristics

On the surface of the lithosphere, the whole hydrosphere and the troposphere below the atmosphere, organisms including animals, plants and microbes constitute a huge and complex ecosystem, namely the biosphere of the earth. There are not less than 500,000 species of animals, not less than one million species of plants and currently uncountable microbes in the ecosystem. In the whole evolution history of the earth, they have been actively participating in the change and development of lithosphere, hydrosphere and atmosphere. Specific geographical environments have specific biocoenoses and specific biogeographical characteristics and nurture various ornamental organisms, forming a variety of peculiar biology tourism resources.

In the evolution history of the earth, biological evolution is an important part of the earth's history. A large number of palaeontological fossils conserved in the strata have become important tourism resources today, such as the dinosaur fossils and dinosaur fossil museum in Zigong City, Sichuan Province, and palaeontological fossil coenosis known as "Rolls of the Book" in Linqu County, Shandong Province. In the evolution of palaeogeographic environments, the living fossils once endangered in geological time such as giant panda, ginkgo and metasequoia have finally survived under specific conditions, becoming important ornamental animals and plants.

In the evolution process of geology, topography, climate and environment of the earth surface since Holocene, a number of unique biological communities have been formed under particular natural conditions and constituted important biological tourism resources, such as the snake community on Dalian Snake Island and the bird community on Bird Island in Qinghai Lake.

Organisms are bound up with natural geographical environment. Particularly, green plants, known as "green transformers", actively participate in the circulation of materials in the atmosphere through photosynthesis, which has an important effect on the physical and mental health of human beings. Plant landscaping has become an important environmental indicator of all modern cities and an important part of all tourist areas, such as pines on Mount Huangshan, spruces on Mount Tianmu and the tropical monsoon forest on Mount Dinghu.

Various nature reserves, which are established in certain natural geographical environments and biogeographic regions to protect wild animals and plants, have become important sites for scientific research and tourism, such as the wildlife reserve in Africa, Shennongjia Nature Reserve in Hubei and Xishuangbanna Tropical Rain Forest Nature Reserve in Yunnan, China.

3.3 Classification of Causes of Formation of Natural Tourism Resources

3.3.1 Brief Introduction to Classification Schemes of Natural Tourism Resources

In the research on tourism science and tourism geography and the preparation of survey and development schemes of regional tourism resources, all scholars classify tourism resources into two categories, namely natural tourism resources and cultural tourism resources. With regard to the classification of natural tourism resources, a variety of classification principles and programs are put forward, which fall into five types:

1. The classification of natural tourism resources by natural elements: tourism geological resources, tourism geomorphic resources, tourism hydrological resources, tourism climatic resources, tourism botanical resources, tourism zoological resources and tourism astronomical resources, or scenic geology, scenic mountains, scenic waters and scenic organisms.
2. The classification of all natural tourism resources mainly by natural geographical environments: mountain-type tourism resources, plateau-type tourism resources, glacier-type tourism resources, grassland-type tourism resources, plain-type tourism resources, desert-type tourism resources, island-type tourism resources, coast-type tourism resources, lake-type tourism resources and river-type tourism resources.
3. The classification of natural tourism resources by tourism functions: sightseeing type, summer resort type, winter resort type, recuperation type, entertainment type, adventure type and scientific expedition type.
4. The classification of natural tourism resources by the distribution pattern and combination of scenic spots in tourist areas: gathering type, radiant type, scattering type, single-line type, circular type and multiple-ring type. For example, the patterns of scenic spots of Five Famous Mountains are summed up as follows: sitting Eastern Mount Tai, lying Central Mount Song, standing Western Mount Hua, walking Northern Mount Heng and flying Southern Mount Hen.
5. The classification of natural tourism resources by frequency of occurrence: common type, ordinary type, special type and singular type.

3.3.2 Classification of Causes of Formation of Natural Tourism Resources

In human history, tourism appeared when science, culture and productivity developed to a certain level, the society was stable and people lived in plenty. Improvement and development of social productivity, science and culture lead to expanding scope and objects of tourism and rising interest and appreciation ability. The purposes of tourism have been shifting from religion/superstition-themed travel to science/culture-themed travel, from sports travel for relaxation and fitness to travel for a change in living environment in order to avoid heat, cold and pollution and from venture travel to scientific expedition. The scope of tourism extends from inside one's own country to outside it, from land to sea and from the earth to the outer space. The means of travel develop from foot and sedan chair to horse and bike, from automobile, train and ship to airplane and airship, and finally to space shuttle and spacecraft today. Natural tourism resources also keep expanding.

In view of the above analysis, the classification of natural tourism resources should be highly generalized, scientific and forward-looking, and should have a clear and in-depth hierarchy rather than give equal weight to large and small, outlines and details, so that the classification scheme may serve as a guide for the exploration and development of natural tourism resources in the future. According to the above guideline and principle, natural tourism resources can be classified into five major categories, 15 categories and 66 landscape types (Table 3.1), with the relationship shown in Fig. 3.1.

3.4 Tourism Resources of Lithosphere

3.4.1 Concept of Tourism Resources of Lithosphere

Man, a member of the earth's biosphere, lives and grows on the interface of atmosphere, lithosphere and hydrosphere. The scenic wonders and mysteries inside the lithosphere and the diversified shapes and charms outside the lithosphere are major natural tourism resources for people to appreciate, research and enjoy. The former are geological natural tourism resources, and the latter are geomorphic natural tourism resources. There are also cave tourism resources in the rocks and formations between the two and at a certain depth below the earth's surface. The three categories of sceneries form tourism resources of lithosphere.

3.4.2 Geological Tourism Resources

The earth's evolution process for more than 4 billion years is also a complex evolution history of geological structure, including differentiation of lithosphere, formation of continental and oceanic crusts, convergence, proliferation, separation and drift of crustal blocks, accompanied by complex orogeny, magmatism, metamorphism and mineralization. Evolution on the surface of the crust includes formation of the atmosphere and hydrosphere, evolution of ancient geographic environment, changes of oceans and lands, appearance of organic matter and living things, formation of the biosphere and appearance of man. The Earth's long and complex history was recorded in the "book" of lithosphere. Chapters of this book include biological evolution, endogenic geological processes, including orogeny, epeirogeny and magmatism, and exogenic geological processes, including climatical and geographical changes, weathering, erosion, transportation and deposition of rocks on the surface of the crust. They have provided us diversified contents for travel, enjoyment and scientific expedition. The chaotic ancient times can inspire our imaginations and delight us with the profound beauty of nature. They are geological tourism resources, the foundation of all natural tourism resources.

Table 3.1 Classification of causes of formation of natural tourism resources

Major category	Category	Tourism landscape	Examples
Tourism resources of lithosphere	Geological tourism resources	Stratum tourism landscape	Laminiform stratigraphic section
		Palaeontological tourism landscape	Producing places of fossils of paleovertebrates and paleoanthropoids
		Tourism landscape of endogenic geological process	Typical structural features, volcanoes and earthquakes
		Tourism landscape of exogenic geological process	Glacial remains
		Tourism landscape of mineral resources	Modern ores and ancient mining sites
	Geomorphic tourism resources	Tourism landscape of erosive tectonic landform	Fault block mountains
		Tourism landscape of tectonic denudation landform	Intrusion landform and Danxia landform
		Tourism landscape of denudation landform	Karst peak forest and Yardang landform
		Tourism landscape of deposition landform	Dune and Mount Mingsha
	Cave tourism resources	Tourism landscape of Karst caves	Limestone cave
		Tourism landscape of volcanic cave	Basalt cave
		Tourism landscape of other types of caves	Sandstone erosion cavern

(continued)

Table 3.1 (continued)

Major category	Category	Tourism landscape	Examples
Tourism resources of hydrosphere	Marine tourism resources	Beach tourism landscape	Bathing beach
		Island tourism landscape	Continental island and oceanic island
		Tourism landscape of coral reef	Great barrier Reef Marine park
		Tourism landscape of tide and wave	Qiantang river tide
	River tourism resources	Tourism landscape of torrent and creek	Hsiukuluan river
		Waterfall tourism landscape	Karst waterfall and fault waterfall
		Valley tourism landscape	Three gorges on the Yangtze river
		Tourism landscape of artificial river	Grand canal
		Tourism landscape of river delta	Danube delta
	Lake tourism resources	Tourism landscapes of structural fault lakes	Qinghai Lake and Dead Sea
		Tourism landscapes of lagoons	Tai Lake and Qili Sea
		Tourism landscapes of oxbow lakes	Hong Lake in Jianghan plain
		Tourism landscapes of glacial lakes	Mahu Lake in Luohuo, Sichuan
		Tourism landscapes of deflation lakes	Juyan Lake and Crescent Lake
		Tourism landscapes of Karst lakes	Grass Sea in Guizhou and Sword pond in Yunnan
		Tourism landscapes of dammed lakes	Diexi Lake in Sichuan
		Tourism landscapes of artificial lakes	Qiandao Lake in Zhejiang
		Tourism landscapes of volcanic lakes	Heaven Lake of Mount Changbai
	Glacier tourism resources	Tourism landscapes of polar glaciers and subpolar glaciers	Glaciers in Greenland, Iceland and Alaska
		Tourism landscapes of temperate glaciers and tropical alpine glaciers	Waterton-Glacier Park

(continued)

Table 3.1 (continued)

Major category	Category	Tourism landscape	Examples
Tourism resources of hydrosphere	Groundwater tourism resources	Tourism landscapes of springs	Quanzhou in Jinan
		Tourism landscapes of hot springs	Yellowstone National Park of the United States
		Tourism landscapes of underground rivers	Underground rivers in Longgong, Guizhou
		Tourism landscapes of mud volcanos and springs	Mud volcanos in Kaohsiung city, Taiwan
		Karez tourism landscapes	Turban basin in Xinjiang
		Tourism landscapes of dragon eyes	The east coast of Jin county, Liaoning Province
		Sinter tourism landscapes	Geyserite stone forest in Tibet
		Tourism landscapes of ancient wells	"Zhen Concubine well" in the Forbidden city
Tourism resources of biosphere	Bantanic tourism resources	Forest landscapes and forest parks	Izmir forest park in Moscow
		Plant nature reserves	Mount Tianmu in Zhejiang and Mount Dinghu in Guangdong
		Botanical gardens	Royal Botanic gardens, Kew
		Flower tourism landscapes	Flower show in Guangzhou and peony show
		Plant attractions in tourist areas	Cypress, the king of trees and exotic flowers and plants
Biosphere tourism resource	Zoological tourism resources	Tourist areas featuring special animal community	Snake Island and Bird Island
		Wildlife reserves	Sichuan Wolong Panda reserve area
		Zoo (Animal museum)	Zoo and aquarium
		Pet tourism resource	Animals for entertainment, relaxation and sports

(continued)

Table 3.1 (continued)

Major category	Category	Tourism landscape	Examples
Tourism resources of atmosphere	Meteorological tourism resources	Polar light tourism landscapes	Northern Europe, North America and Alaska
		Tourism landscapes of Buddha's halo	Mount Emei and Harz mountains in Germany
		Mirage tourism landscapes	Penglai city, Shandong Province
		Tourism landscape of cloud and fog	Sea of clouds in Mount Huang
		Tourism landscapes of misty rain	Nanhu district, Jiaxing City
		Snow tourism landscapes	Snow scene, rime and icicle
		Tourism landscapes of rosy clouds	Sunrise glow and sunset glow
	Climatological tourism resources	Summering climate	Mount Lu, Beidaihe and Iceland
		Wintering climate	Sub-tropical Hainan Island
		Sunshine tourism resources	Coastal sunshine and polar white night
	Clean air tourism resources	Absolutely clean air	Qinghai–Tibet Plateau and poles
		Comparatively clean air	Sparsely populated mountains and coasts
		Forest clean air	Virgin forests and forest parks
Universe tourism resources	Tourism resources of universe and outer space	Tourism resources of outer space	Space shuttle and spacecraft
		Star tourism resources	The Moon and the Venus
	Astronomical tourism resources	Astronomical observation tourism landscape	Eclipse, lunar eclipse and solar physics
		Tourism landscape of visitors from outer space	Meteorites, cosmic dust and cosmic rays

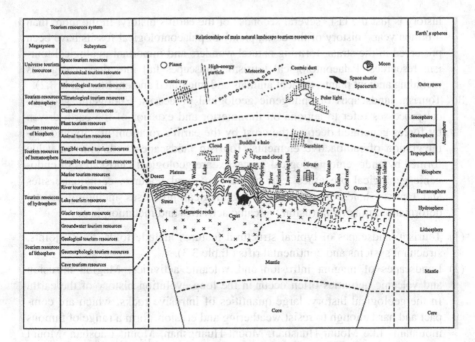

Fig. 3.1 Mode chart of classification of causes of formation of natural tourism resources

1. Stratum tourism landscapes: Stratum generally refers to stratiform rocks and deposits, like sedimentary rocks, volcanic rocks and most of metamorphic rocks. It was formed in the geological time from Archaeozoic era to Cenozoic era and is the fundamental basis for the study of the history of earth. The strata formed in various geological times record the geological events in the process of the earth's evolution including paleoclimate, palaeoenvironment, palaeogeomorphology, ancient life, paleoseismicity, palaeohydrology, ancient tectonic movements, palaeomagnetism and ancient mineralization, all of which can be used for us to make scientific expedition and observe geologic phenomena, learn earthscientific knowledge and explore the mysteries of geological history. Major stratum tourism landscapes include global stratigraphic sections worldwide and regional standard stratigraphic sections. For example, the Middle–Late Proterozoic standard stratigraphic section in the so-called "Ten-thousand-volume Book", that is Jixian County, has been designated as a national nature reserve and receives lots of geological tourists every year. There you can learn and observe the ancient geographical environment, evolution of ancient life, ancient tectonic movements, ancient vulcanicity and ancient mineralization in North China 0.6–1.7 billion years ago and build a scientific concept about the evolution of earth at that time. Moreover, you can collect some ornamental algal fossils, and rocks weathered to different degrees which can be used for rock bonsai.

2. Palaeontological tourism landscapes: The history of the earth is divided by the evolution stage of life into Cryptozoic Eon and Phanerozoic Eon, with the latter further divided into Palaeozoic era, Mesozoic era and Cenozoic era. Human

history is just the last "several seconds" of the earth's history. In the more than 4 billion years' history of the earth, plenty of palaeontological fossils have been preserved in the strata, forming natural wonders and important geological tourism resources. Palaeontological tourism landscapes fall into three categories including ancient plants, ancient animals and integrated ancient life (Table 3.2).

3. Tourism landscapes of endogenic geological processes: Endogenic geological processes refer to actions on the interior and exterior of the earth driven by forces generated deep in the crust by the earth's rotation, gravity and disintegration of radioactive elements. Tourism landscapes with tourism value formed by endogenic geological processes are geological landscapes of endogenic geological processes, e.g. typical structural features, earthquake sites and tourism landscapes formed by geological processes including crustal deformation, magma intrusion, volcanic activities and metamorphism.

(1) Tourism landscapes of typical structural features include folds, faults, joints, structural systems and continental rifts (Table 3.3).

(2) Landscapes of magma intrusion and volcanic activities: Magma intrusion and volcanic activities often occur in the long evolution history of the earth. In the geological history, large quantities of intrusive rocks, which are compact and hard enough to resist weathering and erosion, form a range of famous mountains like Mount Huashan, Mount Huangshan, Mount Laoshan, Mount Hengshan and Mount Wuling. Some small dike intrusive rocks form very spectacular natural tourism landscapes like the roll-like Infatuation Stone in Mount Taishan, which is a result of the wrench movement during intrusion of diabase magma and extends like an earthworm, fascinating tourists.

There are still many active volcanos around the world, with 80 % in three volcanic belts: Pacific Rim, the Mediterranean Sea–Himalayas and the Mid-Atlantic Ridge. Volcano is a geological phenomenon formed by magma erupted out of the earth's crust through a special mechanism. Generally, it refers to the conical highland piled up by erupted materials. During volcanic activity, lots of high-temperature gases, hot water solution, solid debris and lava are ejected. Red-hot active volcanos, dormant volcanos and extinct volcanos are wonders of nature. As a window to detect and study the interior earth and its substances, volcano has attracted many scientific researchers from various disciplines to make investigations and studies and drawn a great many tourists to explore wonders and mysteries of nature.

Lakes, springs, cragged peaks and grotesque rocks formed in volcanic activities provide good natural conditions for tourists to relax and travel. There are many volcanic parks, nature reserves and tourist areas around the world, which can be classified into four types of landscapes: active volcanos, dormant volcanos, extinct volcanos and volcanic cultural relics (Table 3.4).

(3) Landscapes of earthquake relics and crustal deformations: Earthquake is the sudden shake of the earth. Broadly, earthquakes include natural earthquakes due to natural processes and artificial earthquakes resulting from human factors. 80 % of natural earthquakes are caused by crustal tectonic movement. Every year, the world as a whole experiences more than 50,000 felt

Table 3.2 Palaeontological tourism landscapes

Category	Tourism landscapes	Formation conditions	Examples
Ancient plants	Fossil forest, petrified forest, silicified wood, etc.	Silicification and metasomatism of ancient forests and trees buried by mud, sand and volcanic ash under reducing conditions	Fossil forests in Arizona and Utah of the US, silicified woods in Xiadelongwan Village, Yanqing County, Beijing City and silicified wood in Erlong township, Jiang'an County, Sichuan Province
Ancient animals	Palaeontological fossil park, dinosaur museum, museums of australopithecus and paleovertebrates remains	Preservation of skeletons or activity sites of ancient animals after being buried and petrified by sediments	Los Angeles Zoolite Park, Beijing Zhoukoudian Apeman Cave, and Zigong Dashanpu Dinosaur museum
Integrated ancient life	Strata and regions which have preserved plenty of different palaeontological fossils and ruins	Preservation under good conditions after petrification of large quantities of ancient plants and animals which had multiplied and grown in the superior ancient geo-graphical environment	Shandong Linqu mesozoic palaeontologic fossil reserve and museum

Table 3.3 Tourism landscapes of structural features

Tourism landscapes	Formation conditions	Examples
Faults	Divided strata and crust, big and deep faults and fault zone in the upper mantle and recently active faults	Tanlu deep fault in China and San Andreas fault in the US
Folds	Fold or complex fold formed in the stratum by tectonic movement	Large-scale nappe fold in Alps, gravity fold in Mount Song
Joints	Sections formed in tectonic movement, without relative displacement	Vertical granitic joints in Xihai, Mount Huangshan
Structural systems	Structural belts which are composed of structural elements with genetic relations and of different grades, natures and sequences, and complexes made up of blocks	Lotus-like structure in Dalian Mount Baiyun geological park and epsilon-type structure in Malanyu County, Eastern Qing Tombs
Continental rifts	A continental rift, which is formed due to swelling of the mantle and tension of upper crust, may extend thousands of kilometres, forming a spectacular scar on the earth's surface	Great rift valley, Rhine rift valley and Baikal rift

Table 3.4 Volcanic tourism landscapes

Category	Major landscapes	Formation conditions	Examples
Active volcanos	Volcanic vent, magma, fumaroles and hot springs which are being erupted	Joint belt of plate tectonics and volcanic belt of structures like continental rift	Hawaii Volcanoes National Park and Mount Etna in Italy
Dormant volcanos	Volcanic vent lakes, dammed lakes, volcanic landforms and hot springs	Distributed in tectonomagmatic belts, once erupted, now dormant	Heilongjiang five lotus lakes natural reserve and Mount Fuji in Japan
Extinct volcanos	Volcanic landforms like volcanic cone and volcanic vent and hot springs	Volcanos in the geological history and extinct volcanos not erupted for a long time	Volcanic cluster in Datong, Shanxi and Mount Kilimanjaro in Africa
Volcanic cultural remains	Cultural relics preserved after being destroyed by volcanic eruption	Ancient cities and residences buried during volcanic eruption	Vesuvius volcano and Pompell in Italy

earthquakes and 10 strong and destructive earthquakes. Strong earthquakes may change the appearance of nature, forming earthquake remains' landscapes like dammed lakes and fault scarps; strong earthquakes may destroy buildings, forming landscapes of earthquake hazard remains; and strong earthquakes may lead to crustal deformations like vertical and horizontal fractures, dislocations, sandblasting and water leakages on the earth surface. All these, together with lots of steles and sculptures recording earthquakes in human history, constitute tourism landscapes of earthquakes and crustal deformations, whose main categories are shown in Table 3.5.

Table 3.5 Tourism landscape of earthquake relics

Category	Major tourism landscapes	Formation conditions	Examples
Scenic areas of earthquake relics	Ruins of dammed lakes, fault scarps and landslides	Formation of mountains or canyons in middle and high mountainous areas in strong earthquakes	Scenic area of earthquake relics in Xiaonanhai Town, Qianjiang County, Hubei Province
Earthquake hazard remains parks	Ruins of buildings which were destroyed by earthquakes	Deformation, destruction and collapse of buildings in strong earthquakes	Earthquake ruins reserve (park) in Tangshan City
Ruins of crustal deformations	Ground dislocations and twists, fissures, sandblasting and water leakage holes	Mostly formed in the epicentre of fault zones where earthquakes happen	The sunken underwater village in Qiongshan County, Hainan Province, and the earthquake fault dam in Dachang County, Hebei Province
Earthquake steles	Steles recording historic earthquakes	Many stele recordings in earthquake-prone areas	Guangfu temple in Mount Lu, Xichang City, Sichuan Province

(4) Tourism landscapes of exogenic geological processes: Exogenic geological processes refer to actions on the surface of the crust by forces of the air, water and living things generated under the influence of solar energy, gravity energy and lunisolar attraction. Specifically, such processes include weathering, erosion, transportation, sedimentation and diagenesis. Such processes serve as a sculptor of the earth's surface by reducing its undulations and elevation differences. The results of exogenic geological processes are often seen in geomorphic forms. Tourism landscapes of exogenic geological processes mentioned here only refer to typical landscapes with tourism value such as Quaternary glacial remains.

(1) Tourism landscapes of glacial remains: In the evolution history of the earth, paleoclimatic changes led to several great ice ages, including widely recognized Sinian Ice Age, Palaeozoic Ice Age and Quaternary Ice Age. Geological processes of glaciers have left many remains including Sinian moraines in Nantuo, Three Gorges Area, Hubei Province, and Quaternary glacial remains in Mount Lushan, Mount Huangshan, Mount Tianmu and Mount Xishan in Beijing, all of which form important tourism resources. China Quaternary Glacier Vestige Exhibition Hall that is under construction in Moshikou, Beijing, will become an important geological tourism resource in Beijing.

(2) Tourism landscapes of river remains: In the evolution history of geology and topography, river erosion and accumulation processes were in a very complicated manner and some river remains become important natural tourism resources. For example, the yuhua stones (rain flower stones) in Zhonghuamen, Nanjing, were formed by the outwash of the Yangtze River. As for the remains at the pass near the Badaling–Qinglong Bridge, Beijing, China's famous geologist Gao Zhenxi held that they were washed and accumulated by Guishui River, while Professor Feng Jinglan argued in his thesis "Ten Features of the Yellow River" that the ancient Yellow River flowed eastward into Sanggan River from Tuoketuo and then flowed eastward to East China Sea from where the remains are located. Moreover, the river capture remains in Jinsha River, upstream Yangtze River and in the watershed between the White River and Guishui River in the east of Yanqing, Beijing, are also important geological tourism resources.

(3) Tourism landscapes of lake remains: Early Quaternary Nihewan Lake remains in Sanggan River are world famous standard Quaternary stratigraphic section. Upon comprehensive investigations and researches on fish fossils, ancient micro-organisms and ancient man, it is believed that the said remains may be the brackish lake connecting the ocean, near which ancient man lived. Every year, tourists flock here from around the world for scientific expedition and travel.

(4) Landscapes of remains of marine abrasion cliffs and marine accumulation sandbars: Remains of marine abrasion cliffs, which can be found in Dalian, Qinhuangdao, Beidaihe, Shandong Peninsula, coastal areas of Fujian and Guangdong, are important evidences of sea level changes in the geological age and form tourism landscapes; three marine accumulation shell dikes around Tianjin, which provide clues to the time when the local ocean was changed into land, have become important natural tourism resources in

Tianjin thanks to the investigations and researches of Tan Qixiang, a renowned Chinese historical geographer, and the verifications and descriptions of Tianjin Natural History Museum.

(5) Tourism landscapes of mineral resources: Mineral resources refer to available natural mineral resources buried underground (the earth surface inclusive) like metals, non-metal fuels, ground water, inert gases and carbon dioxide. They were mineral deposits produced by a specific geological process under a specific geological environment and used for exploitation and use currently. Man has long been exploiting and using underground mineral resources. Human history can be divided into Stone Age, Bronze Age and Iron Age primarily by natural mineral resources exploited and used. Mines, as geological tourism resource, are on one hand destinations for geological tourism including scientific expeditions. A large number of scientists visit some typical mineral deposits in the world every year, e.g. tungsten mines in Jiangxi, China, saltpetre mines in Chile, copper mines in Zambia, gold mines in South Africa, diamond mines in West Africa and phosphate mines in North Africa. On the other hand, tourists are attracted by some producing areas of precious metals and gem mines and ancient mining relics. In Australia, a 26-ha historical park and a gold mine museum were established in Shu Lau Shan Gold Mining Ruins in the 1850s by reproducing the gold rush in ancient times and have now become a hot tourist spot. In Poland, Salt Crystal Palace—Salt Mine Museum, which was established on the mine 135 m under Wieliczka Salt Mine, attracts more than 0.7 million tourists every year. South Africa has set up a gold mine museum under gold mines. China has also built an Ancient Copper Mining and Metallurgy Ruins Museum in Tongshan Town, Daye City, China.

3.4.3 Geomorphic Tourism Resources

Landforms are a generic term for various shapes on the earth surface. They were formed by endogenous and exogenous processes on the crust. Landform tourism resources refer to landform landscapes with tourism value. According to the intensity and nature of endogenous and exogenous geological processes, landforms can be divided into four major categories, which are erosive structure landforms, tectonic denudation landforms, denudation landforms and accumulation landforms. And according to the nature of exogenous geological processes, landforms are divided into six minor categories, which are river landforms, glacier landforms, Karst landforms, marine landforms, aeolian landforms and gravitational landforms.

1. Erosive structure landforms refer to mountains, cordilleras or mountain ranges formed in the strong uplifts of neotectonic movement. Most mountains in China were formed in this way. For example, the Himalayas, which is the youngest cordillera formed just several million years ago, still ascends by 7 mm per annum. Fault block mountains created in such strong new tectonic movements are the

most magnificent and best mountain tourism resources, as exemplified by most of the mountain tourist areas in East China. For example, the east-west fault in Tazi Village, Shimen Town, which is between Laiwu fault basin developed south of Mount Tai in the Middle Cenozoic era and swelling mountains of Mount Tai faults, recorded 2000 m of differentiated neotectonics, making Mount Tai magnificent and towering. Standing on the denudation surface at the top of Mount Tai, you will feel all other peaks overshadowed by it. Mount Lu is an NNE fault block mountain, with fault cliffs at the east and west towering on the south bank of the Yangtze River. With Poyang Lake on the east and alluvial and aggraded flood plains on the west, Mount Lu looks like a whole range from the side and a single peak from the end. The mountain top is a flat palaeogeomorphoic denudation surface, over 1400 m above sea level. Mount Lu has long been a most famous summer resort in the midstream and downstream reaches of the Yangtze River.

2. Tectonic denudation landforms: They are unique geomorphic tourism landscapes formed by strong erosion in new tectonic movements and under specific strata and tectonic conditions. Such landscapes include (1) Danxia landforms formed by red conglomerates and sandstones and distributed in Danxiashan Scenic Resort in Renhua and Qujiang in northern Guangdong and Hebei Chengde Scenic Resort; (2) cuesta landforms eroded from sloping rock formations, e.g. Sichuan Jianmenguan Scenic Resort and Lingji Peak of Mount Song; (3) rock peak landforms denuded from intrusive rocks and characterized by steep peaks and grotesque stones, e.g. Mount Hua, Mount Huang and Mount Lao; and (4) volcanic rock landforms denuded from volcanic rocks, e.g. Mount Changbai, Devils Tower (basalt orange-like joints), Mount Liuhefang in Nanjing and Mount Tianmu.

3. Denudation landforms: They are tourism landscapes formed by intensive exogenic denudation process under specific strata and structural conditions, including the following: (1) River erosion landforms: river valleys like the Three Gorges of the Yangzte River and Sanmenxia of the Yellow River; meander cores like Mount Xiaogu on the Yangzte River of Pengze, Jiangxi. (2) Glacial abrasion landforms: fish spina/blade-shaped peaks, valleys and cirques, e.g. Mt. Everest, Muztag Ata and Mount Huangshan Fish Spina. (3) Deflation landforms: Yardang landforms formed by strong deflation, e.g. Ghost City in Xinjiang. (4) Loess erosion landforms: landscapes like loess tablelands, ridges, knolls and columns formed by water erosion on extremely thick loess layers, e.g. Mount Baota in Yan'an, Mount Beimang in Luoyang and Yellow River Tourist Area in Zhengzhou. (5) Karst landforms: landscapes formed by soluble rocks or strata (mainly carbonate rocks) strongly corroded by acid rain, surface water and groundwater, e.g. limestone peak forest landforms in Guilin Scenic Resort and stone forest landforms in Yunnan. (6) Marine abrasion landforms: marine abrasion cliffs, capes, terraces, etc. formed by wave erosion, e.g. Ultima Thule Scenic Resort in Ya County, Hainan Province, Beidaihe Dongshan Pigeon Nest and Shandong Chengshanjiao. (7) Gradation denudation landforms: flat plateaus formed by gradation and denudation under exogenous geological processes, e.g. Inner Mongolia Plateau, Guizhou Plateau and Northern Tibet Plateau, which are such special composite tourism landscapes.

4. Accumulation landforms: Landforms formed by materials denuded under various geological processes and deposited and accumulated under certain conditions, e.g. glacial eskers, alluvial fans and river deltas. Some accumulation landforms have great tourism value, like coastal dune landscapes in Changli Golden Coast Resort of Hebei Province, huge travertine accumulation landscapes in Sichuan Huanglong Jiuzhaigou Scenic Areas and aeolian Mount Mingsha in Gansu and Ningxia.

3.4.4 Cave Tourism Resources

1. Karst cave landscapes: Karst caves are complex empty cave systems formed by corroded carbonate rocks or other soluble salt strata. They are horizontal, tilted, approximately vertical or interconnected in many lines or levels. Some of them are dry caves above groundwater level, some are wet caves with regular water flow, some are water curtain caves with underground falls, and some are underwater caves below groundwater level. Karst caves mostly develop along bedding planes, faults and tectonic fissures and are products of corrosion and erosion by flowing groundwater. Karst caves have very complicated shapes. Hall-like caves formed in the intersection of faults and fissures may contain over 1000 people, while caves formed in the single fissure may be too narrow and small to hold even one person. Multi-level horizontal Karst caves connected by different sizes of vertical and slanting branch caves may form a very complex cave system. Some caves may extend for dozens of kilometres and a cave system may reach hundreds of kilometres in total length. A Karst cave takes shape gradually under the action of groundwater in the following two stages: (1) expansion stage: carbonate rocks collapse under corrosion and erosion by groundwater and the cave gradually expands, and (2) congestion stage: when the groundwater level lowers, the cave dries up; rainwater leaks through rock crevices at the top of the cave, and $CaCO_3$ dissolved in the infiltrating water deposits in the cave, forming chemical sediments like stalactites, stalagmites, stone columns, stone curtains and onyx; and finally, the cave is congested by alluvial deposits and colluvial deposits.

Huge underground Karst caves and Karst cave systems and chemical sediments in caves including the sites of ancient men, ancient lives and their activity traces, and ancient cultural relics constitute important natural tourism resources. Karst cave tourism resources are widely distributed in carbonate rock regions all around the world including China, e.g. Reed Flute Cave in Guilin, Seven Stars Cave in Zhaoqing, Yaolin Wonderland Cave in Tonglu of Zhejiang, Zhanggong Cave in Yixing of Jiangsu, Benxi Water Cave in Liaoning, Yunshui Cave and Shihuang Cave in Fangshan of Beijing, and activity traces of ancient men such as Beijing Zhoukoudian Apeman Cave and Liuzhou Bailian Cave.

2. Lava tube landscapes: During volcanic eruptions, molten magma flows and condenses on the ground and becomes a series of complicated lava tunnels or tubes due to activity of various gases in them. These tunnels or tubes, with strangely shaped lava in them, are of great tourism value. Take, for example, the wonderful lava caves at the foot of Ma'an Mountain south of Haikou, Hainan Island, which crisscross each other like a spider web but have different shapes, some magnificent like a palace and some shaped like pythons, winding, deep and grotesque. Among them, Wolong Cave and Fairy Cave are important tourism resources. Wolong Cave has bright basalt walls and is 10 m in width, 3000 m in length and 7 m in height, capable of holding three cars. Fairy Cave is very special: it has other caves in it, with walls covered with coral-like lava bearing visible flow marks. The largest cave is elliptical, measuring 14.7 m in height and 5800 m^2 in area. Stone tables and four big stone columns support the roof of the cave, and in the west is a forest of variously shaped stones.

3. Other cave landscapes include groundwater subsurface erosion caves in sandstones, structural subsurface erosion caves in non-soluble rocks, differential weathered caves and artificial caves. Some of them have important tourism value, e.g. Fairy Cave in Mount Lushan, Jiangxi and Chaoyang Cave in Mount Taishan, Shandong, which were formed by metamorphic rocks under differential weathering and denudation. Some artificial caves also have great tourism value, e.g. tunnels in Jiaozhuanghu, Shunyi, Beijing, and Nandaran, Baoding City, dug during the War of Resistance against Japan, and tunnels in Yongqing County and Xiongxian County, Hebei Province, dating back to the Song Dynasty (960A.D.–1279A.D.) and Sui Dynasty(581A.D.–617A.D.).

3.5 Tourism Resources of Hydrosphere

3.5.1 Concept of Tourism Resources of Hydrosphere (Wang 1985)

In the formation and evolution history of the earth, there appeared a hydrosphere around the earth surface between the atmosphere and the lithosphere. It consists of oceans, rivers, lakes, swamps, glaciers and groundwater. Its total volume is 1386 million km^3, accounting for 0.12 % of the earth's volume. Oceans cover 70.8 % of the earth's surface and have 1338 million km^3 of water, accounting for 96.5 % of the total volume of the hydrosphere; surface water is 25 million km^3, accounting for 1.8 %; and groundwater is 23 million km^3, accounting for 1.7 % of the total. As the hydrosphere, atmosphere, biosphere and lithosphere are closely connected and interpenetrated, water keeps circulating to different extents under solar radiation heat and physical processes, giving rise to various exogenic geological processes and different motive waterscapes, which form a range of tourism resources of hydrosphere.

3.5.2   *Marine Tourism Resources*

The earth's surface measures 510 million km^2 in total area, with 361 million km^2 from oceans, accounting for 70.8 % of the total area of the earth's surface. The northern hemisphere has more land area, but the land area is still just 39.3 % of the total area of the northern hemisphere, as compared with 19.1 % of land area of the southern hemisphere. Sea water plays a very important role in geographical changes of the earth's surface, especially changes of climate. Different locations of oceans (inland sea, epicontinental sea, marginal sea and ocean), different climates (tropics, temperate and cold zones), coasts (sandy coast, rocky coast, muddy coast and coral reef coast), coast landforms, and marine physical, chemical and biological conditions form marine tourism resources with different functions. The main categories are as follows:

1. Beach tourism landscapes: Coastal landscapes and bathing beaches, referred to as "3S", namely Sea, Sand and Sun, are the most attractive natural tourism resources. The best bathing beaches should have flat beaches, fine sand, gentle tides and waves, mild climate and warm sunshine. They were formed in specific coastal landforms and natural geographical environments, e.g. coasts of the Mediterranean.
2. Island tourism landscapes: Land landscapes surrounded by sea, having important tourism value. By formation cause, such landscapes can be classified into the following: (1) Accumulated islands: They are mainly distributed near sea ports not far offshore, e.g. Caofeidian Island on Bohai Sea and Chongming Island on the Yangtze River. Caofeidian Island has been developed as the Bliss Island Tourist Area. (2) Continental islands: They are islands formed by marine erosion or bulging of continental shelf, e.g. Zhoushan Archipelago, Changshan Archipelago and islands in Gulangyu Tourist Area in Xiamen. (3) Oceanic islands: They are widely distributed in the oceans and formed by structural bulge, volcanic eruption or biological process, e.g. Hawaiian Islands, Fiji Islands and Tonga Islands. Those islands, which are like emeralds scattered on the vast ocean, become tourist resorts of the Pacific Ocean.
3. Tourism landscapes of coral reefs: Coral reefs, which are accumulations of calcareous bodies of corals, are distributed in sea water with moderate salinity in tropical regions. They form huge coral clusters consisting of coral reefs and coral islands in the continental shelf and shallow water. The coral islands, with lush tropical forests on them, are just like pearls dispersed on the broad sea. Coral reefs include barrier reefs, atolls and coral seas. Some beautiful reefs have become famous tourist resorts. For example, the Great Barrier Reef Marine Park in Australia attracts a great number of tourists every year, who sit in specially designed glass-bottomed yachts and heartily enjoy various spectacular scenes.
4. Tide tourism landscapes: When the earth rotates and revolves around the sun, the sea flows and ebbs every day due to the gravitation of the moon and the sun. The rising tide is called "tide" in the day and "night tide" in the night, collectively marine tides. In China, tides have been a tourism resource since

ancient times. The high tide of Qiantangjiang River Estuary in Zhejiang Province is the most famous. As the estuary is shaped like a horn, when the sea water rushes in from the wide mouth to Haining where the estuary narrows down to 3 km, huge waves push one after another and form a high water wall. The waves surge and thunder like 10,000 horses galloping. The maximum tidal range is 8–9 m. More than 1000 people watch the tides here during the high tide period every year.

3.5.3 River Tourism Resources

Rivers are important natural tourism resources of the hydrosphere. Most of the countries with ancient civilizations are closely tied to renowned rivers. Rivers not only give birth to ancient civilizations but also form spectacular and enchanting natural landscapes including countless deep gullies, torrential waterfalls and deep pools under the impact of water flow. These landscapes, together with rich cultural tourism resources, form charming tourism corridors along rivers like the Yangtze River and Yellow River in China, Nile in Africa, and Danube and Rhine in Europe. River tourism landscapes mainly include the following:

1. Torrents and creeks: Mostly originating in upstream mountains, they have riverbeds with large longitudinal slope, rapid torrents and marvellous rolling mountains on both sides, offering themselves as good places for rafting lovers. Examples are Hsiukuluan River in Taiwan and Jinsha River in the upstream Yangtze River.
2. Waterfalls: Waterfalls are water flows pouring down from steep slopes or cliffs on the longitudinal sections of riverbeds. They fall into four types by formation cause: (1) Karst waterfalls: They are waterfalls formed under water corrosion in distribution areas of soluble carbonate rocks. Such waterfalls are often accompanied by caves, forming water curtain caves, e.g. Guizhou Huangguoshu Waterfall and Guangxi Li River Waterfall. (2) Waterfalls of structural rock formation: They are formed by hard rocks when river water passes through rocks differing in hardness. For example, Hukou Waterfall on the Yellow River with 34 m of head drop was formed as hard and thick Triassic sandstones narrow the Yellow River from 250 to 50 m. (3) Volcanic lava waterfalls: They are formed by lava blocking river courses during volcanic eruptions, e.g. waterfall of Heaven Lake of Mount Changbai, Jilin, and Jingpo Lake, Heilongjiang. (4) Waterfalls of landslide, debris flow and glacier: Such waterfalls are formed by sediments blocking river courses under exogenic geological process. Sichuan Diexi Waterfall, for example, was formed by blocking of rivers due to earthquakes and landslide.
3. River valleys: They are the most fascinating tourism landscapes. The narrow river, rapid torrents, baffles and shoals in the midstream, and landforms on both banks are nothing short of a fairyland for tourists. Such landscapes are created by new crustal tectonism and intensive river corrosion. For example, the

Three Georges was formed by an arch cut transversely by the Yangtze River. By shape and width, river valleys are classified into ravines and gorges.

4. River deltas: The place where the river carries large quantities of mud and sand into the sea or lake usually forms a nearly triangular plain, which is known as delta. At the places where major rivers in the world empty into the sea, there are huge, variously shaped delta plains with distinctive natural landscapes, forming special delta tourism landscapes. Examples are the Danube Delta tourist area, which has a wide river course, gentle water flow, flat land, fertile soil, thick reeds and flocking birds, forming a wonderful world of river–lake network, villages, fishing grounds, farms and gardens. The Danube Delta is reputed as a paradise of birds. It is the convergence of five migration routes of migratory birds from Europe, Asia and Africa. Over 300 species of migratory birds gather here, forming a bustling and magnificent scene.

5. Artificial rivers: Artificial rivers include canals, drainage rivers and diversion rivers. Important artificial rivers in the world include Grand Canal in China, Suez Canal in Africa and Panama Canal in America.

3.5.4 Lake Tourism Resources

Lakes, which are bodies of water accumulated on low-lying lands, are very important tourism resources. There are two preconditions for the formation of lakes: lake basin and water. In particular, there must be a lake basin and a water source. No lake will be formed if the water source is very limited, or evaporated or lost. Lakes can be divided into inland lakes and exorheic lakes according to whether they are connected to the sea, and salt lake and freshwater lake by salt content. According to the formation causes, lakes are divided into eight types.

1. Structural lake: A water basin formed by fault blocks settled or depressed due to tectonic movement. Examples of such lakes are Lake Baikal in Russia, Lake Tanganyika in Africa, Dead Sea in Jordan and Qinghai Lake, Poyang Lake, Dianchi Lake and Erhai Lake in China. These lakes are generally accompanied by uplifted fault block mountains, like Poyang Lake and Mount Lushan, Dianchi Lake and West Mountain, Erhai Lake and Cangshan Mountain and Qinghai Lake and Riyue Mountain.

2. Lagoon: It is a kind of beach or shallow bay separated by a sandbar, sand spit and sea wall from the sea. At high tide, the lagoon is connected to the sea and becomes a salt lake. If isolated from the sea, it will become a freshwater lake, e.g. Qilihai in Changli Golden Coast Resort. Other lagoons in river deltas like Tai Lake and West Lake form beautiful scenic areas together with surrounding mountains.

3. Oxbow lake: It is formed when a river meander straightens naturally. Examples are Xuanwu Lake and Mochou Lake in Nanjing, East Lake in Wuhan, Hong Lake in Jianghan Plain and Hongze Lake in the downstream of the Huaihe River.

4. Glacial lake: It is a lake formed by water accumulated in low-lying land resulting from glacial erosion or deposition, e.g. Mahu Lake (cirque lake) in Luhuo, Sichuan, and Lake of Heaven (moraine lake) in the northern slope of Bogda Peak in Xinjiang, which make beautiful pictures in contrast to icebergs in the distance.
5. Deflation lake: It is a lake formed by water accumulated in low-lying land created by strong wind action or in a depression in dunes. Its water source is river water or groundwater. Most of such lakes are intermittent lakes or wandering lakes, e.g. Lop Nor in Xinjiang, Juyan Lake in Inner Mongolia and Crescent Lake in Dunhuang, Gansu.
6. Karst lake: It is a lake formed by water accumulated in corroded low-lying land or at the bottom of a corroded funnel in a Karst area. Most of such lakes are small, but have great tourism value when combined with distinctive Karst landforms. Examples are Sword Pond in Stone Forest of Yunnan and Grass Sea in Weining, Guizhou.
7. Dammed lake: It is a lake formed by a river course blocked by gravity sediments like volcanic ejecta, landslide or debris flow. Examples include Wudalianchi and Jingpo Lake formed by river courses blocked by volcanic lava, and Diexi Lake in Sichuan formed by a river course blocked by landslide.
8. Artificial lake: It is a reservoir or an artificial reservoir built in a natural low-lying land in scenic areas. Examples are Qiandao Lake in Fuchun River, Zhejiang, Yansai Lake in Qinhuangdao, and Kunming Lake, Longtan Lake and Jinhai Park in Beijing.

3.5.5 Glacier Tourism Resources

Glacier refers to mobile ice formed by accumulated snow. There is 14.9 million km^2 of glacier on the earth, covering 10 % of the land area and 3 % of the entire earth's surface. By the feature and shape of their formation region, glaciers are divided into mountain glaciers, piedmont glaciers, alpine glaciers and continental glaciers. By geographical location, they are divided into polar glaciers, subpolar glaciers, temperate glaciers and tropical glaciers. By climatic conditions for their formation, they are divided into continental glaciers and maritime glaciers. By feature of tourism landscape, they are divided into two types.

1. Polar glaciers and subpolar glaciers: They are located within or around the Antarctic and Arctic Circles and of great value for scientific and exploration travels. Examples are glaciers in Greenland, Iceland, Alaska and George Island of Antarctica.
2. Temperate glaciers and tropical glaciers: Both are alpine glaciers and important tourism resources. Mount Kilimanjaro in Tanzania at the equator is a very famous tourist attraction in Africa. Waterton-Glacier International Peace Park on the Rocky Mountains in North America measures 1 million acres. In the park, there are more than 50 glaciers and 200 lakes interspersing the mountains and valleys. The park is clad in snow in winter and overgrown with green trees and colourful flowers in summer. In China, the glaciers in Bogeda Peak, Xinjiang, and Gongga Mountain, Sichuan, have been developed as glacier tourist areas.

3.5.6 Groundwater Tourism Resources

Groundwater is water buried in crustal rocks. It may be magmatic water from deep underground, crystal water, adsorption water, film water and pore water in rocks and minerals and diagenetic water and metamorphic water exuded under diagenesis and metamorphism, and "meteoric water" falling from the atmosphere and penetrating into the ground. Under the action of solar radiation energy, earth's gravitational energy, geothermal energy and tectonic energy, groundwater participates in the cycle of hydrosphere, forming various natural landscapes and important natural tourism resources (Yuzhen 1983).

1. Spring landscapes: Spring, a natural outcrop of groundwater, is a natural landscape when groundwater wells out of the ground. It has important tourism value. Examples include 72 springs in Jinan, known as City of Springs, and Yushan Spring in Beijing which Emperor Qianlong called "World No. 1 Spring". By hydrodynamic force of welling, springs can be divided into ascending springs and descending springs; by geological conditions, into erosional springs, contact springs, boundary springs, suspended springs, dike springs, fault springs and karst springs; and by feature and function, into geysers, pulsating springs, shouting springs, smile springs, shy springs, fish springs, fire springs, ice springs, milky springs, sweet springs, bitter springs, medical springs and mineral springs.

2. Hot spring landscapes: They are special natural landscapes when hot groundwater and vapour formed by the action of geothermal energy deep in the crust under certain geological and hydrogeological conditions spurts out of the earth's surface. Owing to their special physical properties and chemical components, such springs have great tourism and medical value. The hot spring in Yellowstone National Park of the United States has become a world famous tourism landscape. Hot springs in Tengchong of Yunnan, Yangbajing of Tibet, Mount Xiaotang of Beijing, Mount Huangshan of Anhui, Tanggangzi of Anshan and Huaqing Pool of Lintong in China are also important natural tourism resources.

3. Underground river landscapes: They are currents that flow through underground Karst passages in areas of carbonate rocks and have the principal features of rivers. The underground passage system composed of mainstreams and tributaries of underground rivers is called underground river system. Accessible underground river courses form tourism landscapes of underground rivers. Examples are the underground rivers in Longgong, Guizhou and Benxi Water Cave.

4. Landscapes of mud volcanoes and mud springs: Under certain geological conditions, when groundwater, underground oil and gases are driven by tectonic forces or magma intrusion, lots of mud, rock debris, gases and some groundwater will erupt out of the earth's surface, forming landscapes of mud volcanoes and mud springs that are similar to natural landscapes formed under volcanic processes. Mud volcano cones are generally less than 10 m in height and several dozen metres in diameter. The largest mud volcano is around Baku oil

field, Caspian Sea, Russia, with the cone reaching hundreds of metres in height. There are 10 odd mud volcanos along a 20-km belt in Kaohsiung City, Taiwan. Mud volcanos and springs are not only valuable as tourism resources but have medical value due to various trace elements contained in volcanic mud.

5. Karez landscapes: Groundwater catching and transmission passages that have been in use in Xinjiang for 1000 of years. A karez is made up of vertical well, underground passage, open channel and sandbar. The vertical well is 25–30 m deep or sometimes as deep as 70 m, and the underground passage is 1.5–2 m high, 0.6–0.7 m wide, 2–3 km long or sometimes as long as 10 km or even longer. The spectacular karezes, which were created by Chinese people to develop and use groundwater in drought areas, attract a great number of tourists for investigation and sightseeing every year.

6. Dragon eye landscapes: "Dragon eyes" are ascending springs gushing out of the bottom of the sea, lake or river. Freshwater springs welling out of coastal areas under the sea are known as "sea dragon eyes". For example, along the east coast from Dalian City to Jin County are a series of ascending freshwater springs welling out of the sea bottom, forming "sea dragon eyes" valuable for tourism and water supply in fishery.

7. Sinter landscapes: Sinters are chemical sediments of groundwater and geothermal water and vapour with dissolved minerals and mineral salts in rock caves or on the earth's surface. Common sinters in nature are classified into sulphur flowers, geyserite, travertine, salt sinter and minerals. Among them, travertine is the most common in nature. Due to special shapes and diversified colours, sinters are of great tourism value. For example, the world unique "travertine stone forest" in the hot spring area of Longmaer, Tibet, has geyserite columns as high as 7 m. Baishuitai sinter in Zhongdian, Yunnan, which was formed by accumulated carbonate sediments is 150 m long and 120 m wide, looking like an oyster-white waterfall. It is a rare natural wonder in China.

8. Well landscapes: A well is an artificial outcrop of groundwater. Many famous tourist areas have wells. For example, in "Bitter and Sweet Well in Baiyuntang", one of the 18 scenes of Mount Heng (in the north), two wells are close to each other, but one is bitter and the other is sweet; "Zhen Concubine Well" in the Forbidden City records a sad story; and "Nine Wells" in "nine ponds, nine pools, nine wells and five bends" in Mount Heng (in the south).

3.6 Tourism Resources of Biosphere

3.6.1 Concept of Tourism Resources of Biosphere

Life forms include animals, plants and micro-organisms. It is generally accepted that life on the earth began to appear 2–2.5 billion years ago, and evolves over an extended geological period from lower life forms into advanced ones and moves from the sea to the land until reaching into every corner of the sea, the land and lower atmosphere, forming the biosphere. Life is always the cynosure in any geographical landscape or in any tourist area. Therefore, biological landscapes (particularly varied botanical landscapes), combined with other natural tourism resources and cultural tourism resources, make up respective tourist areas. Except deserts, icebergs and polar tourist areas, a tourist area without botanical landscapes is unimaginable. Tourism resources of biosphere can be classified into plants and animals.

3.6.2 Botanical Tourism Resources

Plants are an essential part of every tourist area. Botanical tourism resources include the following: forest parks themed on plants; nature reserves predominated by plants; botanical gardens dominated by cultivated plants; and plant attractions, ornamental trees, flowers, etc. in tourist areas. (Zhenheng 1985).

1. Forest parks: Forests can regulate the climate, cleanse the air and absorb dust, and offer themselves as "forest bath" with special medical functions. Finding that natural environment can relax pressure, relieve fatigue and improve health, people establish some green tourism bases—forest parks—in the world. In China, many forest farms and forest zones are changed into forest tourist areas, such as Harbin Forest Park, Forest Park of Banbidian in Daxing of Beijing, Shanghai Gongqingtuan Forest Park, Tuyunguan Forest Park of Guiyang and West Hills Forest Park of Kunming. The most famous forest park in the world is Izmir Forest Park in Moscow of Soviet Union. Measuring 1180 ha, it is home to towering ancient trees, forming beautiful landscapes, and is also provided with various facilities including sanatorium (including a one-day sanatorium), a children's park, an arboured court, a strawhat theatre, an open-air cinema, an exhibition hall, a music hall, a place of amusement, restaurants, cafes and a bathing place.

2. Nature reserves predominated by plants: In China, there are over 100 such nature reserves. Some of them are located in the tropics and subtropics, some are in temperate and warm temperate zones, and the rest are in grasslands and deserts. Each nature reserve has its own special natural landscape, ecosystem and rare botanical species and is of great value for scientific research and tourism appreciation. The relation between natural protection and tourism has been well balanced in each nature reserve, bringing about satisfactory social and economic benefits and scientific research achievements. Take, for example, Mountain Tianmu Nature Reserve in Zhejiang, which is well known for its "mountain's beauty and dense forests", Mountain Trees in Tianmu Nature Reserve are tall, large, old, rare and prosperous. The Nature Reserve is home to most higher plants typically found in the middle subtropical forest zones of China, including not only old plants unique to China like wild gingko, but also over three thousand higher plants (more than 800 medical herbs) such as

Chinese stewartia, Ostrya rehderiana chun and Rhododendron fortune. It is not only an important base for scientific research and education but also a famous tourist area.

3. Botanical gardens: Botanical gardens fall into two types: large comprehensive botanical gardens and featured botanical gardens with special plants. Examples of botanical gardens are Royal Botanic Gardens, Kew Gardens, Beijing Botanical Garden and Nanjing Botanical Garden Mem. Sun Yat-Sen. The beautiful Royal Botanical Gardens, Kew, measuring 1.2 km^2 and rimmed by water on three sides, is located on Thames in the west suburb of London. It used to be a royal villa but was turned into a royal botanic garden in 1841. Here, you can find an extensive plantation, a large library, a herbarium and a laboratory, and dozens of large and spacious greenhouses divided into a palm house, a water plants house, a succulent plants house, a tropical water lily house, an orange garden, an Australian gallery, etc. Some tropical and subtropical plants have lived here for nearly 100 years, and all grow vigorously like palm trees of Chile, cocoa, coffees and bamboos of Mexico, papyruses and various insectivorous plants of Nile, agaves of Latin America, orchids and various cactuses of Brazil, Kangaroo paws of Australia and cypresses of Kashmir. Kew gardens attract millions of visitors every year. Botanical gardens with special plants include botanical gardens reflecting climatic zones like Bogor Botanic Gardens in Indonesia, Xishuangbanna Tropical Botanical Garden, Hangzhou Subtropical Botanical Garden, Nanjing Subtropical Botanical Garden and botanical gardens in temperate zones and cold zones, as well as botanical gardens with featured species like camellia gardens, begonia gardens, succulent botanical gardens, rose gardens, peony gardens, etc.

4. Flower landscapes: Appreciating flowers, growing flowers and holding flower shows take the fancy of a large number of tourists. These activities, often large in extent, have a very long tradition in China, Japan and South-east Asian countries. For example, the annual sukura show in Japan, the flower show of Guangzhou in March, the peony show in Luoyang in April, the chrysanthemum show in Beijing in October, etc. have attracted many tourists. Especially in recent years, the annual peony show in Luoyang has become the grandest tourism project in Central China.

5. Botanical scenic spots in tourist areas: Plants, if having distinctive and peculiar features, can make up an independent scenic spot. Such features are the following: (1) ancient trees, such as Ancient cypress in Yellow Emperor's Mausoleum of Shaanxi, Zhou cypress in Jin Temple of Taiyuan and Sui munifa in Shangqing Temple of Tiantai Mountain, Zhejiang; (2) pretty shape, such as "Guest Greeting Pine" of Mount Huangshan, "Nine Dragon Pine Tree" in Jietai Temple of Beijing, "Love Tree" in the Forbidden City of Beijing, "Love Tree" in Confucius Mansion, "Acacia" in Mountain Tianmu, etc.; (3) beautiful colour, such as "Red Leaves on the Western Hills" in Beijing, "Maple Leaves on Xian Bridge" in Eastern Dongting Mountain, "Ju Yong in Green" in Ju Yong Pass, etc.; (4) attractive blossom, such as "Osmanthus blossom in mid-autumn" in Flower Path of Mount Lu, "Drunk with the Flower" (wood chrysanthemum) in the mountainous areas of Tanzania, "Drunk with the Grass" in Ethiopia,

etc.; (5) large trunk, such as "King of Trees" in Mountain Tianmu, "the Large Ginkgo" in the Qingcheng Mountain, "General Cypress" in Mountain Song, etc.; (6) strange functions, such as bread-producing trees, oil-producing trees, insectivorous trees, etc.; and (7) odd appearance, such as "Chinese Scholar Tree Embracing a Cypress" in Yao Temple of Linfen, three-layered fruit trees in Shaanyin County of Shanxi, trinity tree in Hefeng County of Hubei, etc.

3.6.3 Zoological Tourism Resources

As an essential part of tourism resource system, animals feature various tourism functions including entertainment, appreciation, hunting, fishing, scientific investigation and sustained ecological balance. Main zoological tourism resources comprise the following four categories:

1. Tourism landscapes featuring special animal community: The formation of the landscape of special animal community is related to the evolution of regional geographical environment. Such communities include single animal communities and comprehensive animal communities. Examples of the former are the following: the offshore Snake Island of Dalian, Bird Island of Qinghai Lake, Turtle Island on the Pacific Ocean, Butterfly Valley of Taiwan, etc. An example of the latter is New Zealand, which preserves the fauna and features of the Upper Mesozoic, becoming one of places with the oldest fauna on the earth. Another example is Madagascar, which was separated from the African continent 100 million years ago, forming an independent and special natural eco-environment with animal species, which are extinct in other areas, attracting a lot of scientific workers and tourists.

2. Wildlife reserves: Wildlife reserves include comprehensive wildlife reserves and reserves of single species of important wild animals. All these wild animals are results of specific natural conditions and changing natural environments and are now under priority protection so as to maintain natural ecological balance. An example of the former is the Wildlife Reserve in Nairobi of Kenya, which measures over 200 km in circumference and preserves the natural animal communities typical of tropical grasslands, and tourists can ride in a car and appreciate wildlife in their natural habitat; examples of the latter are found in China, where there are more than 10 reserves of single species of important wild animals, the most famous being Sichuan Wolong Panda Reserve, Giant Salamander Reserve in Jing'an County of Jiangxi, Bawangling Nomascus Hainanus Reserve in Changjiang County of Hainan, etc.

3. Zoos: Zoos include comprehensive zoos and specialized zoos. Their tourism functions are appreciation, scientific research, taming and performance, protective breeding, etc. Currently, there are 887 zoos in 104 countries. Large comprehensive zoos include Beijing Zoo, Ueno Zoo in Tokyo of Japan, London Zoo, Bronx Zoo in New York, etc., which receive hundreds of thousands of tourists every year. Specialized zoos include aquariums, insect museums, bird gardens, butterfly gardens, snake farms, monkey shows and crocodile farms.

Table 3.6 List of tourist activities involving animals

Type of tourist activities	Programs of tourist activities	Main animals used	Example
Entertainment	Circus performance	Horses, monkeys, bears, dogs, pandas, elephants, lions and tigers	Circus in the world
	Routine performance	Elephants, seals, dolphins, crocodiles and snakes	Hong Kong Ocean Park, Thailand Crocodile Park
	Horse racing	Horses	Countries all over the world
Sport	Playing polo and elephant ball	Horses and elephants	Europe and South-east Asia
	Bull fight and ram fight	Bulls and rams	Spain and China
	Hurting	Rabbits, deer, jackals, boars, birds, etc.	All countries
Appreciation, relaxation and fun	Animals for appreciation	Gold fish, tropic fish and small birds	China, Europe and America
	Fishing	Fish	Countries all over the world
	Fun of fighting	Cocks, crickets and golden dogs	China, European and Asian countries
	Listening to bird's singing	Parrots, starlings and orioluses	China and Japan

4. Animal resources for entertainment, appreciation and sports: People raise and tame animals as tools for shows and sports, in order to attract and serve tourists. Major activities involving animals are listed in Table 3.6.

3.7 Tourism Resources of Atmosphere

3.7.1 Concept of Tourism Resources of Atmosphere

The atmosphere encloses the earth, with the surface of water and land as its lower interface. Its upper interface is not obvious. The air 2000–3000 km above the earth is very similar to the interstellar space in density and can be taken as the upper interface of the atmosphere. According to differences of physical properties including temperature and density, the atmosphere can be divided into troposphere, stratosphere, mesosphere, thermosphere (ionosphere) and exosphere. The total mass of the atmosphere is 5.6×10^{19} g, 4/5 of which gathers below 100 km above the ground, and half of the said 4/5 mass is below 10 km above the ground.

The lowest part of the atmosphere is troposphere, which is the closest to the earth's surface and human beings but is significantly affected by the earth's surface. The thickness of the troposphere varies by latitude, season and landform; it is 17 km thick at the equator, 4 km at the poles and 10.5 km at the middle latitude and is thicker in summer than in winter. Different effects of solar radiation, X rays and ultraviolet rays on different places on the earth's surface lead to difference in temperature, density and pressure of the atmosphere, which in turn leads to convection and circulation of the air, giving rise to such weather conditions and processes as cold, heat, dryness, dampness, wind, cloud, rain, snow, frost, fog, thunder and lightning, which form the meteorological and climatological tourism resources with distinctive sights and functions at different places on the earth.

The earth's atmosphere is a mixture of many gases as well as vapour and solid particles contained in the air. The clean air in the earth's atmosphere mainly consists of nitrogen and oxygen, which jointly account for 99 % plus of the clean air. At present, air pollution is becoming more and more serious due to mass production activities and constant population concentration in cities. Therefore, the clean air and environment in nature have also become important tourism resources. The tourism resources of the atmosphere are composed of the meteorological environment, the climatological environment and various clean air environments.

3.7.2 Meteorological Tourism Resources

Meteorology is a generic term for various physical states and phenomena including cold, heat, dryness, dampness, wind, cloud, rain, snow, frost, fog, thunder and lightning. The changing meteorological landscapes combined with the tourism landscapes of the lithosphere, hydrosphere and biosphere, supplemented by cultural tourism resources, form a set of fantastic, natural tourism landscapes. The main meteorological tourism resources fall into seven types.

1. The aurora landscape: Light is the result of the sun's charged particles acted upon by the magnetic field of the earth when they are transmitted to the range of the magnetic field. These particles entering the earth's upper atmosphere from high latitudes activate particles in the upper atmosphere to give out light. The colourful aurora, taking on different shapes, sometimes moving and sometimes still, is the brightest at 5–10 km above the ground, by far overshadowing our festive fireworks. In the northern hemisphere of 22°–27° from the earth's magnetic pole, there is a ring-shaped "auroral zone", generally covering northern Alaska, northern Canada, southern Iceland, northern Norway, southern Novaya Zemlya and southern Novosibirskiye Islands, where people can see the aurora about 245 days in a year, and the aurora becomes one of the main local natural tourist attractions. Aurora also appears once in year in Mohe of Heilongjiang and Altai of Xinjiang.
2. Buddha's Halo (or Ratnaprabha) landscape: Such a landscape usually appears in middle-/low-latitude regions or in clouds on mountain tops. It is a result of slanting sunlight diffracted by cloud droplets and dewdrops. Buddha's halo takes on a luminous ring and moves with people, creating a sense of being in the fairyland. It has become an important meteorological tourist attraction in

mountainous areas. Buddha's halo can be seen in Mount Harz of Germany, Beilugen Mountain of Switzerland, and Mount Lu, Mount Tai and Mount Emei of China, but the Buddha's Halo in Golden Peak of Mount Emei overshadows others. Because there are as many as 320 foggy days on the Golden Peak in a year, averaging 8.9 foggy days and 7.4 rainy days every 10 days, so Buddha's halo can often been seen on Golden Peak.

3. "Mirage" landscape: Mirage is a kind of refraction of light in the atmosphere and happens under special geographical environment and meteorological conditions. It usually appears in Arctic regions in spring and winter, and over deserts, seaside, rivers and lakes in summer. The "Mirage" in Penglai of Shandong is the most famous. With an illusory castle in the air, it is reputed as "Penglai Cloudland".

4. Landscape of cloud and fog: Cloud and fog in mountainous areas change constantly, looking like rolling waves, enough to arouse surging emotions in tourists. For example, "Seas of Clouds" is one of the four wonders of natural landscapes in Mount Huangshan, and "Cloud and Fog" is one of the three wonders of natural landscapes in Mount Lu.

5. "Misty rain" landscape: Gentle misty rain often creates an atmosphere of hazy, poetic romance in the hearts of tourists. "Misty Rain in Nanhu" of Jiaxing is known as one of the scenic wonders south of Yangtze River, and "Watching Rain in the Cloud" of Jigong Mount of Jigong tourist area provides another kind of fun.

6. "Snow" landscape: Snow landscape is an important meteorological tourism resource in middle-latitude areas in winter, especially in mountainous areas, such as "Snow Mantle of Zhongnan" in Xi'an, "Sunny Shaoshi Mountain After Snow" in Mount Songshan and "Sunny Western Hills After Snow" in Beijing. In the ice-locked and snow-swirling northern country, snow landscapes, ice-hanging landscapes and snow sports have become important tourism resources.

7. Rosy clouds landscape: It is a reflection of sunlight passing through clouds at sunrise and sunset. The beautiful rosy clouds, accompanied by sunrise and sunset, form a beautiful picture. They are one of the important natural landscapes in tourist areas.

3.7.3 Climatological Tourism Resources

Climate is the weather in an area averaged over a long period of time and determined by the interaction of sun radiation, atmospheric circulation and ground conditions. The effects of weather in each tourist area on human bodies form the climatological tourism resources in the area. The comfortable or uncomfortable climatological conditions are important considerations of tourists when choosing destinations and dates of tourism. Climatological tourism resources fall into the following types.

1. Summering climate: Summering cities and tourist areas in the world can be divided into three types. (1) High mountain and plateau: This type follows the vertical variety law of climate. Baguio of Philippines, for example, is 1600 m above sea level, with average temperature 6.7 °C lower than Manila. Mount Lu,

a summer resort in China, is 1500 m higher than Jiujiang City, with the average temperature 5.6 °C lower than Jiujiang. (2) Seaside: Owing to the effects of the ocean, the temperature of seaside is lower than that of inland in summer, featuring warmness and moistness, such as Dalian, Qingdao and Beidaihe. (3) High latitude: This type features temperature variation by latitude. Hammerfest of Norway, for example, located at 74°41′ north and facing Barents Sea, is one of the summer resorts in Europe for its comfortable climate in summer.

2. Wintering climate: All winter resorts are located in tropical and subtropical sea climatic zones, such as Hainan Island in China and Sarawak in South Asia. Climate difference between the southern hemisphere and the northern hemisphere is also a consideration in planning our travels. For example, people on the northern hemisphere can go wintering in Australia in Oceania.

3. Sunny climate: Sunlight is an important climatological tourism resource. Countries along the Mediterranean Sea build bathing beaches by taking advantage of the Mediterranean type of subtropical climate, which features long sunshine duration and warm sunshine, and have become well-known tourism resorts in the world.

4. Polar "white night": This is also a kind of sunshine tourism resources. "White night" is called polar day in geography. Owing to the law of the earth's rotation and revolution, the entire Arctic region is exposed to the sun all day long around the summer solstice, forming the "white night" landscape. Summer in the north of northern Europe features mild and moist sea climate, and "white night" lasts for more than 70 days. Northern Europeans who have just got over the long winter nights may find the mild sunshine especially gracious. In inland Europe, summer is intolerably hot, so northern Europe is a good destination for inland continentals to avoid the summer heat and also appreciate the "white night" landscape; therefore, countries in northern Europe receive tens of thousands of tourists in summer every year.

3.7.4 Clean Air Tourism Resources

While the human society is building an increasingly wonderful material civilization for themselves, their living environment is deteriorating constantly because of their activities. Harmful substances generated during industrial and agricultural production and other economic activities seriously pollute the air, water and soil. People who live in poor environments like industrial areas and large cities are in bad need to take a rest in beautiful, unpolluted places when they are off work or on vacation. Therefore, clean air is an important natural tourism resource. It includes three types.

1. Absolutely clean air: It is totally in natural condition without any pollution by man, as on the Qinghai–Tibet Plateau in China, where the air is absolutely clean because of a sparse population and less intensive industrial and agricultural activities. Nowadays, the base sample of the composition of the earth's air is collected here. Cans of clean air from high mountains are now being sold in some countries with serious air pollution.

2. Comparatively clean air: It refers to the air in comparatively clean areas between two or several areas with low air quality. For instance, the air in the Xiaowutai–Jundu mountain area connecting Beijing, Zhangjiakou and Datong is cleaner than that in Beijing City and neighbouring areas.
3. Forest clean air: Forests can cleanse air, adsorb dust, regulate climate, muffle noise and absorb harmful gases like carbon dioxide and sulphur dioxide, and serve to improve pollution. Therefore, it is necessary to generate forest clean air by planting plenty of trees in urban areas and setting forest parks in suburban areas so as to create a quality environment for rest and travel.

3.8 Universe Tourism Resources

3.8.1 Concept of Universe Tourism Resources

Achievements of modern astronomy prove that the earth, where man lives and multiplies, is an ordinary planet in the solar system, the solar system is an ordinary stellar system in the Milky Way Galaxy, and the Milky Way Galaxy and other galaxies compose the metagalaxy. Although the metagalaxy is enormous, it is still very small in the universe. So the universe is infinite both in space and in time, and it is a general term for everything distributed and changing in space. Since the 1950s, with the rapid development of modern science and technology, man has reached from the earth into the outer space in the universe and landed on the moon for the first time, and is now striving to land on Venus by the end of the twentieth century. Therefore, man's interest in tourism has also extended to the universe from the earth. Astronomical visions including outer space, celestial bodies and extraterrestrials make up the universe tourism resources.

3.8.2 Outer Space Tourism Resources

In the early 1960s, Yuri Gagarin, a Soviet citizen, made a trip to the space outside the earth by spaceship for the first time, viewing the beautiful scenes of the earth and the moon. In the early 1980s, the crown prince of Saudi Arabia travelled to the outer space at his own expense. Scientists from various countries made multiple scientific experimental trips to the outer space by spaceship and space shuttle. Space Adventures Corporation in Seattle of the US decided to organize a group to tour the outer space in 1993, with the itinerary being 4–5 circles around the earth along the space orbit, lasting 8–12 h. The shuttle named "Space Exploration" of the Corporation is under construction. Up to now, 250 people have booked the tour at a price of $52,220, which includes expenses of the tour and two space meals. Tourists can view the earth and the outer space from the windows of the shuttle, and the guide will explain matters relating to the outer space.

3.8.3 Star Tourism Resources

Man landed on the moon in the 1960s and is now conducting scientific explorations in preparation for landing on Venus. Some countries with advanced science are working out plans to exploit mineral resources on the moon. It will be soon for man to tour the moon, and celestial bodies including the moon will be tourist areas in the future.

3.8.4 Astronomical Observation Tourism Resources

As the earth is part of the huge universe, changes of extraterrestrial bodies directly affect the earthly environment where man lives, such as solar radiation, magnetic storm, sunspot activities and cosmic rays. Man started astronomical observations and records as early as 5000 years ago. Astronomical observations become a tourism resource because of man's great interest in astronomical observations and phenomena. For example, Ancient Observatory in Beijing, Dengfeng Observatory and modern Purple Mountain Observatory in Nanjing, Beijing Observatory and Beijing Planetarium aiming to disseminate astronomic and scientific knowledge attract thousands of tourists every year. Greenwich Observatory, which is near London, UK, is a world-class tourist attraction.

In addition, many tourists are also interested in observing annular eclipse, total solar eclipse, lunar eclipse, oppositions of Venus and Mars, and comets. In the observation of Halley's Comet in 1986, for example, Australia, as the best observation zone, attracted many astronomers and amateur astronomers to come here for observation and tourism. Another example is the path of total solar eclipse along Taiyuan, Handan and Nantong in 1987, which also attracted plenty of foreign astronomers and amateur astronomers. The opposition of Mars, which happens only once every 32 years, took place in late September 1988, and tourists, astronomers and amateur astronomers came to the Xikamaka Observatory in Peru due to its good observation conditions.

3.8.5 Meteorite Tourism Resources

Meteorites, known as "visitors from outer space", had been regarded as the only samples of excelestial bodies before man brought back lunar rocks. It is estimated that there are 1500 meteorites weighing with the mass over 100 kg getting near the earth every year. However, their mass is no more than 10 kg when they fall to the ground, so only 4–5 pieces are detected per year, but some huge meteorites still can be found. Currently, more than 1700 types of meteorites are collected and preserved in the world. They fall into three categories: stony meteorites, iron meteorites and stony-iron meteorite. There are dozens of meteorite museums in the world, which attract many tourists every year and have become important tourism resources for scientific research and dissemination of astronomic knowledge. In China, Meteorite Museum in Jilin and an iron meteorite weighing 30 tons in Xinjiang are both important tourist attractions.

References

Geography Department of Nanjing University. (1963). *Geomorphology*. Beijing: People's Education Press.

Kalesnik, C. B. (1947). Основьг общегоЗемлеведения, Учцедтиз [1955. Principles of general geography] (T. Yongluan, Trans.). Beijing: Higher Education Press.

Wang, Y. (1985). *Amusing things in geography*. Lanzhou: Gansu People's Publishing House.

Writing group of tourism introduction. (1983). *Introduction to tourism*. Tianjin: Tianjin People's Publishing House.

Yunting, L. (1988). Modern tourism geography. Nanjing: Jiangsu People's Publishing House.

Yuzhen, C. (1983). *Introduction to Chinese historic interests*. Beijing: China Travel & Tourism Press.

Zhengzhou Geology University. (1979). *Geomorphology and quatemary geology*. Beijing: The Geological Publishing House.

Zhenheng, J. (1985). *View of famous spots overseas*. Beijing: Popular Science Press.

Chapter 4
Earthscientific Formation Conditions of Cultural Tourism Resources

Although cultural tourism resources are a result of man's cultural activities over time, their formation and distribution depend largely on natural environments and such factors as history, nations and ideology. Therefore, the territorial rules of the generation, development and distribution of cultural tourism resources are also research objects of tourism earth-science workers by virtue of the principles and methods of geology and geography. In this book, we just analyse the conditions in China.

4.1 Earthscientific Characteristics of Historical and Cultural Tourism Resources

4.1.1 Locality of Historical Cultures

Any historical culture emerges and exists in a specific area. Cultures formed in the same area but different ages are different, and cultures formed in the same age but different areas show greater difference. For example, in the Qing Dynasty (1616AD–1911AD), Beijing was the capital, the national political and cultural centre as well as the regular place of residence of emperors. To facilitate emperors to have fun and avoid summer heat, royal gardens were built in or near Beijing rather than in Guangzhou, but Guangzhou, with its proximity to the South China Sea and excellent harbour and Pearl River waterway, was sure to become an important port for maritime trade in the south. Thus, Guangzhou took the lead in commerce and culture, and owing to constant exchanges with the outside world, tended to have a broad mind for new things. Towards the end of the Qing Dynasty, Guangzhou became the key base for Sun Yat-sen to lead the Bourgeois Revolution. Obviously, the remarkable differences between the type and nature of cultural

© Springer-Verlag Berlin Heidelberg and Science Press Ltd. 2015
A. Chen et al., *The Principles of Geotourism*, Springer Geography,
DOI 10.1007/978-3-662-46697-1_4

tourism resources of Beijing and Guangzhou are ascribable to cultural and natural conditions.

Geographical environments have significant effects on historical cultures. Generally speaking, southern China is humid with beautiful natural landscapes, while northern China is dry and cold and predominated by plateaus and large plains. Compared with southern China, northern China has less thriving vegetations on mountains and smaller water volumes in rivers and has suffered frequent natural disasters in history. Therefore, the different natural environments in southern and northern China are sure to lead to two styles of cultures, differing in music, dance, drawing, poetry and folk customs. In respect of drama in particular, the Cantonese (Yue) Opera of the South, with its beautiful aria and delicate performance skills, excels at expressing feelings and presenting the flavour of waterfront cities in southern China; by contrast, the Shaanxi Opera (Qinqiang) of the North, with its strong and heroic aria and simple and wild performance, is suggestive of the grand Loess Plateau. From the perspective of the history, martial spirit is admired in the North, while literature is valued in the South, which has a lot to do with specific natural environments. It is the regional nature of historical culture that forms the earthscientific characteristics of cultural tourism resources.

4.1.2 Earthscientific Characteristics of Cultural Sites of the Stone Age

For a long period of time after "Australopithecus" evolved into man two to three million years ago, man could not make bronze tools but had to fight for a living in nature with "stone tools" made by striking and grinding. Archaeologists referred to this phase of man's history as the Stone Age and divided the Stone Age into Old Stone Age, Mesolithic Age and New Stone Age according to the workmanship of stone tools. Anthropologists correspondingly divided the development stages of man into ape man, Homo neanderthalensis, and homo sapiens by brain size, intelligence and physical shape. In geological history, the Stone Age consists of Early Pleistocene, Medio Pleistocene and Late Pleistocene of Quaternary and the Holocene Epoch.

1. *Regional nature of cultural sites in the Old Stone Age*
   Regional distribution of cultural sites in the Old Stone Age So far, the oldest site ever found in China is the Site of Yuanmou Man, followed by Sites of Shaanxi Lantian Man and Peking Man. Some scholars hold that, owing to climatological evolution, like the northward shift of climate zone in the glacial period of the Quaternary, man moved to the north from the south, which is in line with the argument that the ancestors of man are the "Australopithecus" living in the tropical forests. In addition, sites of ancient man radiate southwards, northwards and westwards from its centre in the middle and lower reaches of the Yellow River (including Haihe River). For example, of the 22 cultural sites

of the Early Old Stone Age found up to now, 15 are located in the Yellow River
Basin, accounting for 68.2 % of the total; except one site located in northern
Guangdong, the other cultural sites of the Middle Old Stone Age are distributed along the Yellow and Yangtze Rivers; and the cultural sites of the Late Old
Stone Age are found throughout China, as in north-east China, east China, central south China and south-west China. However, the Yellow River Basin still
takes the lead in the number of sites, indicating that the Yellow River Basin was
a favourable living environment for man as early as tens or hundreds of thousands of years ago.

Environmental characteristics of cultural sites in the Old Stone Age Most
cultural sites in the Old Stone Age, except some located at river terraces and
hillocks, are natural caves, especially in the Middle and Late Old Stone Age.
Modern Tourism Geography written by Lu Yunting lists nine caverns of ancient
man, namely Jinniu Cave in Yingkou of Liaoning, Stone Dragon Head Cave
in Daye of Hubei, Meipu Cave in Yunyang of Hubei, Bodhisattvas Cave in
Qianxi of Guizhou, Dove Cave in Kazuo of Liaoning, Xiaonanhai Cave in
Anyang of Henan and Turtle Cave in Jiande of Zhejiang. There are also some
caves where ancient man once lived, such as Tongtian Cave (Liujiang Man)
in Liujiang of Guangxi, Bailian Cave in the suburb of Liuzhou, Chenjia Cave
and Siduo Cavern in Liujiang County and Lion Cave (Maba Man) in Maba of
Guangdong. Caves where ancient man lived are mostly limestone caverns, for
such caverns often have spacious "halls". The limestone caverns where ancient
man lived had basically stopped developing, and many of them had been lifted
to a certain height due to crustal movements so that they were inundated by
floods. These caverns were dry inside. To breathe fresh air, ancient man often
lived near the mouths of the caverns. Such caverns are mostly near water
sources, with their mouths sunwards and leeward (to avoid cold winter winds),
with none northwards or north-westwards.

Lithology and rock structure in the Old Stone Age Stone tools of ancient
man were made of hard rocks taken near where they lived. Scrapers, chopping
tools and pointed tools used by Lantian Man, for example, were all made of
quartz rock, vein quartz and quartz sandstone. In limestone regions, all tools
were all made of limestone and dolomite. Dingcun Man made chopping tools
with hornfels collected nearby, which is flaky and very hard. In Late Old Stone
Age, stone tools, mainly scrapers and pointed tools, were mostly made from
slates and processed on one side.

2. *Regional nature of cultural sites in the New Stone Age*
   Selection of cultural sites in the New Stone Age As communities and villages
   of clan societies of primitive communes, cultural sites in the New Stone Ages
   feature the following. (1) Proximity to water sources for easy access to water,
   e.g. Banpo Site in Xi'an, which is located at the left bank of the Yellow River;
   Peiligang Historical Site, which lies at a turn of Wei River (Shuangpo River);
   Yangshao Site, which faces waters on three sides and mountain on one side;
   and Dawenkou Site, which lies on both sides of Dawen River; (2) high altitude

to facilitate water drainage and avoid flooding. More than 400 Yangshao Sites have been found in Guanzhong (the Wei River Basin), and many of them are located on terraces of Wei River and its primary and secondary tributaries; 13 sites have been found on tablelands and hilly areas for 40 km along the Feng River (Chen 1983). Near the sites were limited areas of flat land for farming purpose; (3) proximity to rivers, sunward and leeward: Cultural sites are most in sunward and leeward places on the northern and western banks to avoid cold winds and facilitate daylighting.

Rules of distribution of cultural sites in the New Stone Age To date, 6000–7000 cultural sites in the New Stone Age have been found throughout China. Their distribution follows the following rules. (1) Highest density on the plains in the middle and lower reaches of the Yellow River, including Guanzhong Plain, plains in the lower reaches of Fen River in south-west Shanxi, plains in the lower reaches of Yiluo River, plains on the two banks of Ji River in north-east Henan and west Shandong. There are over 900 Yangshao Sites found within Shaanxi; (2) second highest density on the plains in the middle and lower reaches of Yangtze River, including Jianghan Plain and the valleys on the Three Gorges of the Yangtze River, Hangjiahu Plain in Taihu Basin and areas between the Yangtze River and Huai River. Dozens of sites found in Zhejiang are mainly centralized in the lower reaches of Qiantang River; (3) centring on the middle and lower reaches of the Yellow River, and stretching outward in descending density from river valleys and mountain foot. For example, distribution of sites extends westwards from the valleys of Wei River to Hexi Corridor and the valleys of the Yellow River in Gansu and Qinghai, and north-eastwards from Lanzhou to Ningxia Plain; northwards along the valleys of Fen River and the foot of Mount Taihang to intermontane basins in northern Shanxi and northern Hebei, and finally to the north-east areas; and eastwards to the seaside; (4) small density and age in south and south-west China.

Cultures in the New Stone Age characterized by primitive agriculture are concentrated on the plains in the middle and lower reaches of the Yellow River, which is determined by the superior natural conditions described as follows (Ma et al. 1987). (1) Flat terrain, suitable for farming: Guanzhong Plain, 300–600 m above sea level and known as 800-li Qinchuan, boasts flat terrace and loess tablelands mixed by plateaus (loess plateau) and marshlands on both banks of the river valley. The tablelands are wide and flat and safe from blooding and are therefore suitable for farming. Most regions east of Mount Taihang and Mount Song and between the Yellow River and Ji River are part of North China Plain less than 50 m above sea level. Distributed on the plain are microhighlands along ancient rivers, namely "Qiu (hillock)", such as Shangqiu, Diqiu, Taoqiu and Renqiu. Such hillocks have mild gradient which facilitates drainage and are therefore favourable for farming; (2) many rivers and lakes, facilitating irrigation and water transportation. Guanzhong is where Wei River meets Jing River and Beiluo River, and to its east are Fen River, Yiluo River and Qin River. The Yellow River crosses this area, but a change in its course resulted in the formation of some interchannel low-lying lands, and low-lying lands at the front of and between alluvial fans at the foot of mountains, where water gathered to form lakes. At that time, marshes like Daluze, Dayeze, Heze and Mengzhuze and Shenze and ancient lakes including Ningjinpo Lake on North China Plain stored flood water in the flood season and provided water in the dry season; with abundant aquatic animals and plants, they became one of the food sources for ancient people; (3) warm and moist climate, favourable for growing crops and living. Research findings indicate that from the Middle New Stone Age 5000–6000 years ago to early Western Zhou Dynasty (1046BC–771BC) 3000 years ago, the Yellow River Basin featured a subtropical climate and was much warmer and moister than today. Skeletons of wild animals including roes, bamboo rats and racoon dogs unearthed at Xi'an's Banpo sites all belonged to subtropical animals. Roes now live in the marshlands of the Yangtze River Basin, and bamboo rats feed on bamboo shoots and roots. The shift of their habitats to the south of Qinling Mountain indicates that the climate of the Yellow River Basin then was similar to the subtropical climate today; (4) fertile soil, suitable for agriculture. Most of the Yellow River Basin is covered by alluvial and aeolian silt loess, which is loose and porous and good for the growth of crops; (5) thick forests and grasslands in the Yellow River Basin in the Middle and Late New Stone Age, with diversified plants and wild animals, providing rich resources for ancient residents to cultivate crops and raise livestock.

4.1.3 Earthscientific Characteristics of Cultures in China's History

It is generally accepted that the written history started from the Xia Dynasty (2070BC–1600BC) that was founded in the twenty-first century BC. Some legends before the Xia Dynasty broadly corresponded to the Later New Stone Age. With the improvement of productivity, the development of the society, the replacement of dynasties, and the transfer and dissemination of culture, different areas feature different cultures. Undoubtedly, these different historical cultures influenced the formation and distribution of some cultural tourism resources.

1. *Early cultural centre—middle and lower reaches of the Yellow River*
   The middle and lower reaches of the Yellow River are an important cradle of China's ancient civilization. As already stated, man lived and multiplied here as early as in the Stone Age. For thousands of years thereafter, many dynasties set their capitals here in ancient cities including Anyang, Xi'an, Luoyang, Kaifeng and Shangqiu, which, together with towns and villages nearby, have preserved plenty of cultural relics and places of interest, such as Yin ruins in Anyang; Emperor Qin Shihuang's Mausoleum, Mao Mausoleum of the Han Dynasty (202BC–220AD), Qian Mausoleum and Zhao Mausoleum of the Tang Dynasty (618AD–907AD), and Ci'en Temple with Big Wild Goose

Pagoda and Jianfu Temple with Small Wild Goose Pagoda (built at the prime time of Buddhism) in Xi'an; White Horse Temple, graves of the Han Dynasty (202BC–220AD) and Longmen Grottoes (lit. Dragon's Gate Grottoes) in Luoyang; Dragon Pavilion, Iron Pagoda and Po Pagoda in Kaifeng; capital site of early Shang Dynasty (1600BC–1046BC) in Yanshi; and ancient city wall discovered in Zhengzhou.

The distribution of the above historical and cultural sites is tied up with historical activities of human beings and natural environment. Take Xi'an for example. Xi'an is flanked by Qinling Mountain in the south, North Mountain in the north and Long Mountain in the west and has the Yellow River to the east. In addition, the Wei River crosses the middle of Guanzhong Plain and "eight rivers" nearby run around Xi'an. These mountains and rivers provide "Xingsheng" (superior geographical location) for Xi'an. Xi'an has "natural barriers at four sides": Wuguan to the south-east guarding the way to Jing and Xiang; Dasan Barrier to the west leading to the centre of Shu; Gold Lock Pass to the north fending off the Huns; and Tong Guan and the valleys of the Yellow River lying to the east (Shi 1981), which makes Xi'an easy to attack others and defend itself, fending off the Huns from the west and controlling the Central Plains to the east. It is easy for Xi'an as capital to handle matters relating to border defence. Furthermore, with fertile soil, developed agriculture, rich products and hydraulic projects started in the Qin and Han Dynasties, the Guanzhong Plain had long been reputed as "Land of Abundance". According to Sima Qian's records in the Western Han Dynasty (202BC–8AD), "The land area of Guanzhong occupies 1/3 of the total area of the country, but the population here just accounts for 30 % of the total and the output accounts for 60 % of the total." (See *Historical Records • Merchant Biography*). Another example is Luoyang, capital of nine dynasties. Leaning against Mangshan Mountain and looking towards Yique, it has the world's best Xingsheng, with natural barriers to buffer attacks from invaders and the Yellow River leading to rich regions, avoiding the Sanmen Natural Barrier in water transport. Despite a small area of flat land, Luoyang has its geographical advantages. All the dynasties taking Luoyang as capital built walls and palaces, none of which have been preserved. Another example is Kaifeng, one of the seven ancient capitals in China, which is located at the edge of North China Plain. It has flat land and a dense river network. The Yellow River, man-made Bian Canal, Cai Canal and Guangji Canal greatly facilitated water transport of grain from south of the Yangtze River. In this respect, Kaifeng excelled Luoyang and Xi'an. What is described above shows that the natural environment played a vital role in the relocation of capital from the west to the east and was also one of the basic conditions for the distribution and change of cultural tourism resources.

2. *The development of regions south of the Yangtze River and the shift of the economic focus to the south*

Although primitive agriculture appeared in some areas south of the Yangtze River in the New Stone Age, the economy and culture of these areas lagged behind the Yellow River Basin after the Qin and Han Dynasties. However, with natural conditions like water and heat superior to those of the north and supported by stable farm yields with fewer disasters, regions south of the Yangtze River are provided with favourable conditions for economic development. On the contrary, the Yellow River Basin experienced three large civil disturbances and the southward march of northern minority peoples, which gave rise to massive migration of people from the Central Plains. Meanwhile, agricultural production technologies were brought to the south, where agriculture and water conservancy experienced great development. In the middle of the Tang Dynasty, the south overtook the north in economy, and in the Song Dynasty (960AD–1279AD), there was a saying "A good harvest in Suzhou and Huzhou makes the world rich". Economic development in the regions south of the Yangtze River contributed to cultural prosperity, as described in the old saying "Jiangsu and Zhejiang are a gathering place of fortune and talent" (Chen 1983). Generations of celebrities and beautiful landscapes south of the Yangtze River fostered a "delicate" flavour featuring tender music, beautiful dance, exquisite buildings and fine embroidery, which are typical of south of the Yangtze River. Most places of interest reflecting the economy and cultures south of the Yangtze River have rich regional characteristics, such as ancient city walls and tombs, Mochou Lake and Xuanwu Lake in Nanjing, Temple of Soul's Retreat and West Lake in Hangzhou, and gardens and temples in Suzhou and Wuxi.

3. *Beijing—cultural centre in the modern history of China*
   In the late feudal era of China, Jin, Yuan, Ming and Qing Dynasties all made their capital in Beijing, lasting for over 700 years. Beijing's rise as the national centre of politics and culture is ascribable to its Xingsheng, long history and superior strategic position. Beijing leans against Yanshan Mountain in the north, is lined by the Sea on the left, embraces Taihang Mountain on the right and approaches the estuaries of the Yellow River and Ji River in the south. According to ancient people, "With the world's best Xingsheng, flanked by mountains and lined by sea, plus solid city walls and deep moat…Beijing is really a capital for endless dynasties." (Yu et al. 1985). Canyons in the Yanshan Mountain and Taihang Mountain lead to Bashang, the area north of the Great Wall and Liaodong. In the south lies the North China Plain. It can be said that "Beijing controls the Yangtze River and Huai River in the south and is within easy reach of the north". Beijing stands on the ancient alluvial fan of Yongding River, whose ancient river courses and low-lying lands formed lakes including "Six Lakes". Haidian District in the western suburb has landscapes typical of southern Chinese riverside cities, and the beautiful woods and springs of Xishan Mountain make Beijing top others in mountain springs. In terms of strategic location, Beijing is the only road from the Inner Mongolia Plateau and north-east China to. The Liao Dynasty (907AD–1125AD), the Jin Dynasty (1115AD–1234AD) and the Yuan Dynasty (1206AD–1368AD) took Beijing as the foothold to advance to the Central Plains and each built Nanjing City,

Central City and Da Du City in Beijing. A lot of historic relics preserved in Beijing are mainly related to the events that the Liao, Jin, Yuan, Ming and Qing Dynasties made their capital here. Such historic sites include Tianning Temple Tower built by the Liao Kingdom, Lugou Bridge built by the Jin Kingdom, the Grand Canal dug by the Yuan Dynasty, the Forbidden City, altars and temples, mausoleums built by the Ming Dynasty (1368AD–1644AD), and western suburban gardens, a large number of temples and palaces built by the Qing Dynasty.

4.2 Earthscientific Characteristics of Ancient Architectures and Projects

Ancient architectures like palaces, pavilions, alcoves, platforms and bridges, and ancient architectural works like the Great Wall, moats, dams, sea walls and canals all belong to historical architectural culture featuring remarkable forms of material culture (Kuang 1987). Over years of vicissitudes, these ancient architectures either were reduced to ruins or remain intact today. By observing and studying them, people can not only understand the economic and technological developments as well as cultural and art styles in a certain historical period, but also find out the close relations between their locations, capacities, building materials, hues and styles and earth-science, for any ancient architecture or work has its own specific site condition and environment. Even in natural scenic areas, it is required that all architectural landscapes should harmoniously develop with natural landscapes. Ancient architectures and works fall into many types and differ not only in shape, but also in size and capacity. Quite a few of them are historical and cultural heritages valuable to China and the whole world and have been listed as important historic relics under state protection. We will study the earthscientific characteristics of ancient architectures by the following categories: wood and stone architectures, garden architectures, ancient mausoleums and ancient architectural works.

4.2.1 Wood and Stone Architectures

By building materials, architectures can generally be divided into timber architectures, earthwork architectures, brick and stone architectures and wood and stone architectures. In particular, wood and stone architectures are the most common and the most typical traditional architectures which demonstrate earthscientific characteristics.

1. *Timber architectures*

   Timber architectures refer to architectures (like towers, pavilions, palaces and halls) whose support frameworks were made of timber. They were created by our ancestors in a specific environment. In ancient times, the climate was warm and humid and forests thrived both in the middle and lower reaches of the Yellow River and in the Yangtze River Basin and Huai River Basin, so it was very easy to get, cut, transport and process timber. Excelling in bearing capacity and integrity, timber is suitable for architectures of various sizes on different landforms. Therefore, timber architectures become the main tradition of Chinese architectures.

Chinese timber architectures date far back from Banpo Man in Xi'an and Hemudu Man in Yuyao of Zhejiang 6000–7000 years ago to the Ming and Qing Dynasties and up to the modern times. Although timber architectures have kept improving in structure and art, they have generally maintained the tradition that is typical of the Chinese architectural system. To sum up, they have the following earthscientific characteristics:

The flexibility in adapting to regional differences Timber architectures can appear singly or in groups. In both cases, their capacities and architectural styles may adapt to local conditions. For example, columns used in architectures of different functions may differ in number, capacity, shape and structure. Generally, old pavilions are supported by four, six or eight columns; large pavilions are built with massive columns; for example, the roof of Hall of Supreme Harmony in the Forbidden City of Beijing is supported by 72 wood columns, which are carved with pictures of dragons and phoenixes or are just left unadorned. The layout of doors and windows is very flexible. Because walls do not bear weight, the number, sizes and positions of doors and windows are determined by functions and local conditions. For example, pavilions and corridors may dispense with windows; living rooms may have back windows for ventilation purpose; and anterooms may have large windows for daylighting. Group architectures place more emphasis on layout art to demonstrate art beauty and variation beauty of space. China is a vast, multinational country, so the layout of architectural groups differs from place to place. However, the common feature of Chinese timber architectures lies in symmetry of plane layout, namely centring on the median line of the main architecture, with the left part and the right part or the front part and the back part arranged symmetrically. Additionally, the orientation of this layout is compatible with the natural environment, giving people a sense of symmetric beauty.

Compatibility between architectures and environment Ancient Chinese architectures are good at using the natural environment to create an inductive atmosphere to set off the functions of architectural groups. This was effectuated in two ways. One was to create the inductive atmosphere through an ingenious layout. For example, in front of the Hall of Supreme Harmony in the Forbidden City of Beijing is an empty lot, which measures 30,230 m^2 but has no tree on it. Officials about to have an audience with the emperor had to across the empty lot and then step up into the Hall. They were sure to feel overshadowed by the majestic Hall of Supreme Harmony, and then, the image of the emperor would appear sovereign in their minds. The other way was to take advantage of terrain and other natural conditions to build architectures with different capacities, shapes and structures. Architectures on hillsides would be built to suit the terrain. For example, the Eight Outer Temples in Chengde, mostly built at the foot of mountains, look very grand and solemn. The Yellow Crane Tower in Wuhan built on the Snake Mountain and Aiwan Pavilion in Yuelu located in the valley demonstrate the principle that "Towers emphasize height and wilderness and Pavilions on quietness and depth". Bridges were built over streams and pavilions near water, with water circling and roads winding. All these are typical examples of happy marriage between ancient Chinese architectures and environments.

Superiority in preventing sunstroke and dispelling cold Ancient Chinese people were good at using different shapes and structures of timber architectures to prevent sunstroke and dispel cold. For example, they used brackets to support beams and tie beams and long eaves, because such long eaves served to protect people from direct sunshine and rain; for another example, they used brackets and wooden frames to form large roof, which provided a large room between the indoor top and the roof to keep off cold in winter and heat in summer. These are the notable attributes of timber architectures.

Integrity in withstanding earthquakes and sinking In earthquake resistance, timber architectures far excel masonry structures with load-bearing walls. This is because pillars, purlins, beams and tie beams of timber architectures are all connected with mortises and tenons, brackets and slanting props, which can tightly engage them as a whole. Even if one part is damaged, the whole architecture will not collapse. In particular, the elasticity of the whole framework system provides some flexibility for the mortise–tenon structures. Facts show that in case of an earthquake, only the guarding walls collapse, while the whole wooden framework remains intact or are slightly damaged. The Wooden Pagoda in Ying County of Shanxi, built in the second year (1056) of Qingning's reign of the Liao Dynasty, has survived several earthquakes. The well-known Dule Temple in Ji County of Hebei survived the magnitude 8.0 earthquake in Sanhe and Pinggu in 1679, while all the houses in the county seat of Ji were destroyed. In the magnitude 7.8 earthquake in Tangshan in 1976, the white pagoda built in the Liao Dynasty in front of the Dule Temple was damaged and some small buildings built in the Ming and Qing Dynasties also suffered serious damages, but the Avalokitesvara Hall was safe. The stone tablet, the holy temple, Xiangfen of Shanxi, recorded that in the earthquakes in the seventh year of Dade's reign of the Yuan Dynasty, "The monks' dormitory, vestibule and kitchen of the temple were ruined, but the hall alone remained intact", which is because the hall had high-standard brackets, but the monks' dormitory, vestibule and kitchen did not have such brackets.

2. *Masonry architectures*

Masonry architectures are an essential part of ancient Chinese architectures, and their engineering technologies have reached a high level over thousands of years, for instance, the famous Great Wall, massive royal mausoleums, exquisite brick and stone pagodas, diversified bridges and grottoes with fine arts.

These masonry architectures are closely linked to geological and geographic environments, and their earthscientific characteristics are as follows.

Architectures relating to lithology Grottoes were often chiselled in the following rock strata: mixed formation of red sand and loess, conglomerate formation, red sandstones, granitic rocks, limestone formation, quartz sandstones and shale formation. In particular, most grottoes were chiselled in sandstone and conglomerate formations, because these rocks are moderate in hardness, resistant to erosion and easy to chisel. However, loess is loose and easy to get weathered and collapse along vertical joints, which is bad for chiselling grottoes; limestone is hard but erodes easily when exposed to underwater or fissure water. Therefore, statues and their preservation status in grottoes vary by lithology. Dunhuang Grottoes, for example, were chiselled in the hard and brittle conglomerate of Yumen System, so frescos and clay figures predominate in the grottoes, while Yungang Grottoes in Datong were chiselled in evenly textured sandstones, so statues predominate in the grottoes. Natural damages also vary from grotto to grotto.

Architectures relating to landforms For example, ancient pagodas were built on different landforms according to different functions. Pagodas for landscaping purpose were built on high mountains or at the turn of rivers. Baochu Pagoda, standing on Baoshi Mountain of West Lake in Hangzhou, symbolizes the beautiful West Lake, and Yufeng Pagoda on Yuquan Mountain of Beijing is an adornment to Xishan Mountain and has become a borrowed scene of the Summer Palace. These brick and stone pagodas add beauty to local mountains and rivers.

Architectures relating to geographical locations Grottoes are generally distributed near ancient transportation and trade roads, and their locations indicate the dissemination path and development level of history and culture; some other ancient pagodas providing distant views and serving to keep watch on enemy activities are often distributed on ancient defence frontiers, and ancient pagodas serving guiding and piloting purposes were mostly built at river and sea ports, at the turns of rivers, and near bridges and ferries, for example Pagoda Anchorage in Mawei of Fuzhou, which was listed as a key marine mark in early times; Liuhe Pagoda in Hangzhou, which was built at a turn of Qiantang River and serving as an important market for ships sailing at night; and Zhenfeng Pagoda in Anqing of Anhui, which stands at a turn of the Yangtze River and has always played a role in "guiding ships with lighted lamps" (Luo 1985).

Architectures relating to earthscientific studies Some ancient pagodas indicate ancient geographical environments and are of scientific value for studies on changes of geographical landforms in history. The White Pagoda at Zhakou of Hangzhou was built in the Five Dynasties and the early Song Dynasty (960AD–1279AD). Studies show that 800–900 years ago the Pagoda once stood at the estuary where the Grand Canal flowed through the city and met Qiantang River, but now it is located on a small hill. Therefore, its positions and surrounding landforms in the past and at present bear witness to the changes of geographical environments of Hangzhou in history.

Architectures relating to resistance to natural disasters For example, Zhejiang-Shanghai Stone Sea wall, reputed as "Great Wall on the sea", was built to guard against the tidal waves of Qiangtang River. To consolidate the sea wall and protect coastal farmlands, ancient people gradually turned the early earth sea wall into the massive stone sea wall.

4.2.2   Garden Architectures

In history, gardens were built as multifunctional landscape complexes for ruling and rich classes to seek fun and pleasure. Their basic characteristics are as follows: made by man but fitting naturally into the environment. Therefore, every garden is a typical artificial landscape with mountains and rivers, as described by the ancient people "Although they were built by men, it seems that they were created by nature". Gardens mainly consist of trees, mountains and rivers, and architectures, with architectures being the main components. In ancient times, construction of gardens, especially private gardens, began with building of architectures, first flower halls and then trees and stones. Apparently, gardens are predominated by architectures and supplemented by trees and stones, which are appendages to architectures. By nature and function, ancient Chinese gardens can be classified into royal gardens, monastery gardens, ancestral gardens, private gardens and gardens with famous mountains and waterscapes. However, all gardens were designed and built with architectural landscapes as the central theme of the spatial art complex. These architectures should have distinctive features in the following three aspects.

1. Habitat excelling not only in natural beauty with fun in the wilderness but also in living beauty with strong local flavour: This, reflected in gardening techniques, may be generalized in the following contents. (1) Different plane layout: Unlike palaces and temples, gardens are laid out flexibly and naturally, without remarkable middle lines (except for small gardens). Take Summer Palace in Beijing for example. Its main entrance is Orient House, and its main architecture is located at the foot of Longevity Hill, facing south and lakes. Another example is Humble Administration Garden in Suzhou, whose overall layout centres on mountains and ponds. All towers, platforms and pavilions are built beside water at different heights and with different styles. (2) Flexible locations: Gardens are built near mountains or rivers. For example, regarding the site selection for Imperial Summer Resort in Chengde, Emperor Kangxi in the Qing Dynasty once said "It is nature itself without artificial work". It maintains the natural landforms, with the low-lying land south of Hot Spring dug to simulate landscapes of southern Chinese riverside cities. Architectures like palaces either hide in mountains or overlook lakes, with large flat terrain (like pear garden) left unchanged. The mountain villa made up by hills, flat terrain and lakes looks like a miniature of China's topography. (3) Winding paths:

presenting natural beauty: "Paths in the garden are like eyebrows and eyes". A garden will be lively with well-designed paths and spiritless without them. Therefore, paths should be winding, to give people room for imagination about what lies ahead. (4) Flowers and trees: They form a natural mix full of fun and sentiment. Take for example willow, a common ornamental tree in gardens. It is seldom seen in gardens south of the Yangtze River. This is because willows need to be planted near water and in rows, and their thick leaves and branches look like draperies in the house but lack elegance, so they are not suitable for small gardens in the south. By contrast, gardens in the north are very large, providing sufficient space for towering willows, which add a kind of tenderness and delicacy to the gardens. Therefore, the arrangement of trees and flowers in gardens should match local conditions, with some hidden and some exposed, all naturally of course.

2. Painting, that is the construction and layout of gardens, are like the scores in music and should be based on artistic selection and generalization of natural landscapes to create infinite imagination in a limited space so that "a small peak suggests a high mountain and a spoon suggests a large body of water". It is just like painting, which is incapable of life-size representation of mountains and rivers. Therefore, gardens should be treated artistically both in terms of overall layout and in terms of static or dynamic layout. For example, in forming the framework and scenery in gardens, all the halls, rooms, houses, alcoves, palaces, pavilions, bridges, doors, corridors, rockeries and ponds are well designed so that "scenery comes out of conception and is appropriately laid out". Halls for receiving guests and handling political affairs are solemn; small-sized alcoves and pavilions, either as landscapes or as their ornaments, are generally delicate; and corridors in gardens serve to link scenes, so they should be long, smooth and winding. In a word, artificial architectures should be flexible and built based on artistic conceptions and landscapes.

Building rockeries in gardens is a time-honoured tradition. Later in history, nearly all gardens have rockeries, some standing in ponds, some on the bank and some in the yards. The artistic conception of rockeries comes from nature, but they are laid based on a high generalization of natural mountains. Rockeries can be strange peaks or hills and may have caves or prominent stones, but all should be treated artistically.

Gardens in China feature outstanding painting conceptions and landscaping techniques. Common techniques are obstructive scenery, enframed scenery, borrowed scenery, vista line and opposite scenery. Obstructive scenery is based on the appreciation psychology of hiding in order to show, with the main features and the most beautiful scenes hidden by trees, rockeries and wall arcades, so that when people tour gardens, they will first view secondary scenes and then primary scenes, gradually reaching the climax and getting the best anaesthetic experience. Enframed scenery refers to external scenes enframed through various doors and windows into a specific perspective range in order to select some scenes and reject the others. For example, mullion designs on windows or spaces between branches and leaves make up vaguely beautiful scenery, namely leaking-through scenery. With regard to borrowed scenery, Chinese gardens are good at "borrowing from others", as reflected in many successful examples like the "Sunshine on Hammer Mountain" of Imperial Summer Resort in Chengde, which takes advantage of the scenery of Hammer Mountain outside the Resort. Scenes in Jichang Garden of Wuxi all face mountains, with external mountain scenery included into the garden.

With respect to spatial treatment, it is often the case that there are gardens in gardens, yards in gardens, islands in lakes and lakes in islands, so that large gardens do not seem void, small gardens do not seem narrow, large lakes do not seem large and small ponds do not seem small. The lakeshore is winding, with knaggy rocks divided by banks to achieve the optimal artistic effects.

In harmonious contrast, movement and stillness, shade and light, emptiness and fullness, and low and high are treated ingeniously. In a word, Chinese gardens, when treated artistically, look like a poetic landscape painting.

3. Artistic conception: This is the essence of Chinese gardens. Artistic conception means delivering a certain inclination of the owner and expressing some conceptions about love and hate or beauty and ugliness, which is the romantic ideal beauty characteristic of Chinese gardens. This artistic conception is usually embodied in landscaping, names, pillars, couplets, inscriptions, flowers and trees. Take for example the Garden of Couple's Retreat in Suzhou. Its name, landscaping and layout all express the beautiful love between Shen Bingcheng and his wife in the Qing Dynasty. The water-themed Master-Of-Nets Garden in Suzhou, for another example, expresses the intention of Song Zongyuan in the Qing Dynasty to "retire for fishing", hence the name "Master of Nets". "Washing My Ribbon Pavilion" in the garden takes its name from the opposite meaning of a fisher's song to Quyuan, "If the water of the Cang Lang River is clean, I wash the ribbon of my hat; if the water of the Cang Lang River is dirty, I wash my feet.", in order to convey the sentiment of beauty. Lotuses in many gardens aim to represent the artistic conception that "Lotus flower is not stained when coming out of mud and remains unpretentious when bathed in clear water". Bamboos, being straight, are planted in gardens to express the idea of straightness, not bending to evil practice.

The poetic and picturesque sentiments of gardens are rooted in their artistic conception. Chinese gardening techniques come down in a continuous line with, and are rooted in, splendid ancient literature, painting and calligraphy.

4.2.3 Ancient Mausoleums

Ancient mausoleums with value of earthscientific tourism include royal mausoleums and mausoleums of celebrities. Both the location and layout of the two kinds of mausoleums are closely linked to the natural environment, and it is necessary to examine their tourism value from the earthscientific perspective.

1. Mausoleums of emperors: Locations of mausoleums of emperors are inseparable from the ancient people's belief in "fengshui". They believed that good fengshui would bring good luck to families. In particular, rich families always try to find places with good fengshui for their mausoleums in order to boost their fortunes. Emperors and royalties, who think they are above others, are pickier about locations of mausoleums. They not only own places with Xingsheng topping the whole country or the whole district but also build very grand mausoleums. Before they died, some emperors forced skilled craftsmen and tens of thousands of labourers all over the country to build extravagantly expensive mausoleums, dreaming that they would continue their monarchical life beyond this world. Therefore, royal mausoleums preserved until today reflect to some extent the architectural style and the artistic level as well as the superior natural and location conditions at the time. Although "fengshui" is superstition, locations of mausoleums objectively demonstrate some superior earthscientific characteristics and add attraction to earthscientific tourism to mausoleums. Some royal mausoleums are described in brief below.

Emperor Qin Shihuang's Mausoleum Located 5 km east of Lintong, it leans against Mount Li in the south and borders on Wei River in the north. South of Mount Li is Lantian, which is reputed as "land of jade" for its rich jade reserve. Such Xingsheng adds charm to the geographical atmosphere of the Emperor Qin Shihuang's mausoleum. Moreover, the luxury of Emperor Qin Shihuang made the mausoleum very grand. In the district once stood a city. The perimeter of the outer city was 6 km, and that of the inner city was 2.5 km. The mausoleum, piled by earth on the flat land of the inner city, measured 46 m in height and 1410 m in perimeter, shaped like a covered funnel. The underground construction works of the mausoleum are even more tremendous. Besides three vaults containing terracotta figures which have been excavated, there are also 17 mortuary tombs, 70 convicts' tombs and 93 horse tombs. With its large-scale, numerous mortuary objects and excellent craftsmanship of terracotta figures, Emperor Qin Shihuang's Mausoleum is regarded as the eighth wonder of the world.

Mausoleums of the Western Han Dynasty Nine mausoleum areas, in which were buried 11 emperors, are distributed on Xianyang Tableland north of Wei River, stretching over a hundred li from the east to the west. The earthscientific characteristic of mausoleums of the Han Dynasty (202BC–220AD) is that these mausoleums are all located on loess tableland, which is suitable for building underground chambers due to its flat but high terrain, thick layer of earth and large clearance above underground water; the open terrain is suitable for building groups of mausoleums so as to enhance its Xingsheng.

Zhaoling Mausoleum of the Tang Dynasty The location selection for mausoleums of the Tang Dynasty also follows the traditional Xingsheng idea of the Qin and Han Dynasties, but outperform the two dynasties by taking better advantage of mountains. Since then, the practice of building mausoleums on mountains has been established. Mausoleums were built in low places of North Mountain in Guanzhong. In plane layout, square walls were built surrounding mountains and hills, with four doors in four directions and watchtowers at four corners. The tomb passage paved in front of the mausoleum climbs up the ramp step by step, and the stone archway as well as the stone human statues and rock beasts tidily arranged at both sides were carved exquisitely and look magnificent. It is noteworthy that in appearance, the tomb takes advantage of the natural peak of Jiuzong Mountain to set off the sovereign dignity and verve of the imperial mausoleum; the chamber built directly at the lower place of the mountain is very stable and can guard against theft. The peaks at the two sides form a natural door of the mausoleum, making the mausoleum more magnificent and spectacular. In a word, the characteristic of Xingsheng-based mausoleums is best embodied in the layout and architecture of Tang mausoleums.

Mausoleums of the Ming and Qing Dynasties The earthscientific characteristic of these mausoleums is also outstanding. Most of them are located at the southerly side of small basins surrounded by mountains. Take for example the Thirteen Tombs of the Ming Emperors, which are located in the small basin at the south foot of Jundu Mountain. To its north, it is under the protection of Heavenly Longevity Mountain, and to its south, it is guarded by Dragon Mountain and Tiger Mountain which face each other. Several small rivers coming from the North Mountain join in the basin, forming the wide tableland in the north of the Thirteen Tombs Basin. The mausoleums of 13 emperors of the Ming Dynasty are arranged along the rim of the basin, forming a semi-ring-shaped mausoleum group. Eastern and Western Tombs of the Qing Dynasty were built in Zunhua and Yi County of Hebei, respectively. The former is located at the rim of an intermontane basin in front of Changrui Mountain. At the south rim of the basin stand Tiantai Mountain and Yandun Mountain, which face each other and form a gate in between. The tombs face south and look very majestic. The latter leans against Yongning Mountain and is surrounded by high cliffs on four sides, with Zijing Pass to the west, Yi River to the south and Langya Mountain to the east. It is a quiet place with beautiful scenery.

2. Mausoleums of celebrities: In addition to imperial mausoleums, there are also a lot of mausoleums of celebrities, including meritorious persons, and famous prime ministers, warriors, literati and scientists. Owing to their outstanding contributions and immortal services, descendants built mausoleums and commemorative temples for them. Examples are Qu Yuan's Mausoleum in Leping Li of Zigui County, Hubei; Zhang Liang's Mausoleum in Lankao County of Henan; Confucius Cemetery in Qufu of Shandong; Zhang Qian's Mausoleum in Chenggu County of Shaanxi; Si Maqian's Mausoleum in Hancheng of Shaanxi; Wen Tianxiang's Mausoleum in Ji'an County of Jiangxi; and Memorial Temple of Lord Bao in Hefei of Anhui. Mausoleums of celebrities are usually located at the hometowns of the deceased, working places when they were alive and places which they once admired, and some of the mausoleums of celebrities were built on the battlefields where they died. Preserved over hundreds of years, these mausoleums have become verdant and scenic destinations. The popularity of celebrities before death combined with the beauty of landscapes adds more colours to the earthscientific tourism here.

4.2.4 Ancient Architectural Works

Ancient architectural works including the Great Wall, moats, hydraulic works, large bridges and sea walls preserved until today not only represent the essence of Chinese history and culture, but also are essential parts of earthscientific tourism resources. Their earthscientific tourism values lie in the following: (1) fully utilizing and adapting to nature, and combining the architectural works and surrounding objects into an integrated geographical entity; (2) determining the locations, capacities, structures and hues of architectures according to the trend and morphological features of mountains and rivers so as to blend the architectures as a harmonious landscape into the surrounding environment; and (3) contributing to improving natural environment, developing regional economy and traffic, and promoting political and cultural exchanges; some of these ancient works are still of great geographical significance.

Several important ancient architectural works:

1. The Great Wall: As an ancient defensive system, the Great Wall first appeared in 700BC in the spring and autumn period and was rebuilt or extended in most dynasties over two thousand years until the fourteenth century when the Ming Dynasty fell. For example, the great walls built by the kingdoms of Chu, Zhao, Qi and Wei in the spring and autumn period, and the great walls built in the Qin Dynasty and the Han Dynasty (202BC–220AD) have also become historical relics. The great walls built in the dynasties of Wei and Jin, Northern and Southern Dynasties, and dynasties of Sui, Tang, Song and Yuan are also basically reduced to ruins and historical relics. Only the Great Wall built in the Ming Dynasty from the bank of Yalu River to Jiayu Pass at the foot of Qilian Mountain is totally preserved. The Great Wall was formerly named "Border Wall" and measures over 6350 km. With the longest construction period, the largest investment and the strongest structure, it is reputed as a "wonder of the world" and has become the "No. 1" and most attractive landscape of historical and cultural tourism in China. The earthscientific values of the Great Wall built in the Ming Dynasty are reflected in the following three aspects:
   Outstanding cultural and geographical landscapes The Great Wall is the most massive architectural landscape on the earth, whose western and middle parts extend basically along the boundary between arid and semi-arid regions. The western part winds westwards at the rim between the Gobi Desert and the plateau; the middle part is based on the northern ridge of Mount Heng or lies on the border between the Inner Mongolia Plateau and the highland; and the eastern part, featuring more complicated terrain, was built along the northern ridge of Mount Yanshan, winding from south to north over mountains and across valleys, with barriers in passes and water towers built on heights. Take for example the building materials of the Great Wall of the Ming Dynasty. It was very dry and seldom rained along the western part of the Great Wall, the building materials for that part were the earth collected locally; the eastern and middle parts, due to the rainy weather in summer and complicated terrain and

because of their position as the key defensive section, were built with materials collected from mountains, with stones laid at the bottom, bricks laid at the top and earth filled and rammed in between, forming a spectacular landscape. Such a long, complete, grand and risky architectural work is unique on the earth. No wonder that someone made such a calculation: if the materials used in the Great Wall of the Ming Dynasty were built into a wall 2 m thick and 4 m high, the wall would be by far longer than enough to circle the Earth; if the materials were built into a road 5 m wide and 35 cm thick, the road could circle the Earth 3–4 times. Lunarnauts also said that from the moon, they could merely see the Great Wall of China and Sea walls of Netherland out of all human architectures on the Earth. The Great Wall of China is therefore called the longest architecture in the world and listed as one of the wonders of the world.

Delicate design and complete system The Great Wall is a masterpiece of ancient defence works in China. For example, on the 95-li section of Jinshanling Great Wall stand 242 watchtowers and battle platforms. Watchtowers fall into three types: wooden structure, brick structure and wooden and stone structure. In appearance, watchtowers are square, rectangular, round or oval; inside the watchtowers, there are single rooms, double rooms, as well as whistle stands, shooting holes, living rooms, secret passages, battle platforms and food storages. In a word, the delicate design and various shapes of watchtowers are really amazing. Every several steps on the wall, there are an internal indentation and external indentation for holding rolling logs and stones. On the top of the wall, there is a 3–5-m-wide connective channel, which can be used by soldiers to support each other at war. Beacon towers set at heights of the Great Wall served to transmit information. Moreover, at dangerous places are 2-m-high barrier walls with steps built along the mountain terrain. Therefore, even if the enemies had climbed upon the Great Wall, they would meet resistance from soldiers on higher vantage points. At some sections, low walls protected by trenches were built at the outer side of the Great Wall as the first defence line. To conclude, all defence works were organically combined and supplemented each other, forming a rigorous and complete defence system.

Boundary mark of natural environment The middle and eastern parts of the Great Wall tally with the 400-mm isohyet and have become an important artificial architectural mark of natural geographical division in China, showing differences in regional economic development in China. South of the Great Wall is basically agricultural zone with good natural conditions; north of the Great Wall is agriculture–pasture zone or pastoral zone featuring arid and semi-arid natural environment. Such opposite landscapes indicate that the location and trend of the Great Wall are by no means determined arbitrarily but chosen after field investigation and careful consideration of various factors including landforms, hydrological conditions, history and economy. As a result, the Great Wall is not only of military strategic significance but also of profound geographical significance, and it represents the landform of a large geological belt.

2. City walls and moats City walls and moats are enclosures built around ancient capitals, local administrative centres, important frontier and key places of commerce and traffic. Moat or trench dug outside the city forms a complete architectural system for military defence. Moats or trenches no longer provide protection today, but as cultural heritages, they are of certain historical significance and appreciation value and of direct significance to the development of tourism.

Most ancient city walls and moats in China have become relics after thousands of years' damages by war, nature and man. However, some city walls and moats remain intact today, such as those of Nanjing, Xi'an, Jingzhou, Xingcheng, Pingyao, Taigu, Dali and Chaozhou. Owing to their different geographical locations, their shapes demonstrate significant regional characteristics.

Large ancient cities and northern city walls and moats Capital cities and ancient cities on the northern plains generally have square or quasi-square city walls, outside which are moats or trenches. A flood wall is often built at the side facing the river, such as the flood wall south of Zhengding City of Hebei and the flood wall west of Miyun City of Beijing. In ancient cities in the north, open channels were dug to introduce river water into the cities, forming landscapes of open channels beside streets.

City walls and moats around southern Chinese riverside cities Ancient cities in the watery regions in South China had city walls built astride water channels to facilitate water transportation. These city walls had many water gates, which was a major feature. For example, the Shanghai City built in the 32nd year (1553) of Jiajing's reign was round and had four water gates.

City walls and moats around mountain and river cities in South China In mountainous regions and valleys in the south, cities are mostly located at terraces or tablelands on the northern or western banks of rivers. For example, Zigui, Wushan, Fengjie and Yunyang all lie on the northern bank of the Three Gorges, and their city gates facing the Yangtze River all look to piers; ancient cities along Minjiang River, which flows from the north to the south, are mainly located on the western bank. Apparently, their locations serve to avoid cold winds and conform to the fengshui idea that facing east will bring good luck. The planar shapes of city walls and moats are not necessarily square but vary according to the layout of the terrains and river turns.

City walls and moats around cities along the Silk Road Ancient cities left on the ancient Silk Road in north-west China are predominantly located on the oasis at the edge of deserts, such as Wuwei, Zhangye, Jiuquan and Dunhuang. As a land transport artery between the east and the west in ancient times, the Hexi Corridor is the only route of the Silk Road, and many cities were built here due to its important geographical location, such as Heishuikou City, Camel City and Loulan Ancient City in the west, all of which are reduced to ruins now. It is ascribable to desertification resulting from climatic changes and excessive reclamation; meanwhile, frequent wars and failure of these cities to serve as defence strongholds or business and transportation centres are also important causes.

City walls and moats around cities on North-east Plain The city walls of some border cities such as Mergen and Aihui on the northern low-lying plains in north-east China are built with peat log, which are dried after being dug from low-lying lands. These materials which are collected locally save labour and are very solid. Bounding walls of government offices and private residences inside cities are all built with two lines of palings with earth filled between them, showing very strong local features.

3. Ancient hydraulic works: Ancient hydraulic works include canals, dams, karezes and sea walls, which are represented by Beijing-Hangzhou Grand Canal and Ling Canal, Dujiang Dam, karezes in Xinjiang and sea walls in Jiangsu and Zhejiang. These hydraulic works are of significant scientific, historical and appreciation values both for earth-science and tourism and play vital roles in the economic development of China. Currently, these ancient hydraulic works have been rebuilt into tourist areas of cultural and architectural landscapes, attracting a lot of tourists from home and abroad. Their earthscientific characteristics are mainly reflected in the following three aspects.

   Outstanding example of creative development of nature The aforesaid works, just like the Great Wall, are the greatest ancient architectural heritages in China and even in the world. They represent the level of highly developed ancient Chinese civilization and are essential parts of the largest existing cultural landscapes on the earth. (1) The Grand Canal: It was first built in the spring and autumn period and the Warring States Period (475BC–221BC) and experienced several changes of courses, links and extensions from the Sui Dynasty (581AD–617AD) to the Yuan Dynasty. Measuring 1794 km, it runs through Haihe River, Yellow River, Huaihe River, Yangtze River and Qiantang River and is the oldest and longest canal in the world. Its prominent functions, profound influence and spectacular architecture overshadow any other cultural heritage in the world. (2) Dujiang Dam: Located at Minjiang River's outlet at the north-west corner of Chengdu Plain, Dujiang Dam was built to prevent flooding by Minjiang River and irrigate the farmlands on Chengdu Plain. The work was finished in 300–250BC and is now over 2000 years old. It irrigated over three million mu of farmland (now eight million mu[1]), making Chengdu Plain "the Land of Abundance", "where people are free from all care about drought or famine". The outstanding construction technology and long-term benefits of Dujiang Dam are also peerless in the world; (3) Ling Canal: Also known as Xing'an Canal or Xianggui Canal, it is located within Xing'an County of Guangxi and is one of the greatest canal works in ancient China. It is connected to Linghe River and Xiangjiang River in the upstream of Lijiang River and links Lingnan to Lingbei. Although it is just over 30 km long, it lies between waters with different base levels and runs through mountains, indicating great difficulty in digging the canal at the time. However, ancient Chinese people wisely used the techniques of dams and locks to build such a

---

[1] 1 mu ≈ 666.7 m^2

mountain-crossing lock-type canal. The well-known poet Fan Chengda in the southern Song Dynasty (1127AD–1279AD) once said "Ling Canal is peerless for its ingenuity in water control". (4) Karezes: They are a great invention for farmland irrigation and agricultural development in arid areas like Hami and Turpan of Xinjiang. During hundreds of years, thousands of hydraulic systems with complete structures including shafts, culverts and outlets were built here, ranging from 3 km to 20–30 km in length. Over 1100 karezes have been preserved, with a total length of more than 3000 km. Its long history and outstanding digging sinking method are marvellous. Even now, karezes are still used by local people, forming a comprehensive cultural and earthscientific landscape unique to Xinjiang.

Outstanding example of careful analysis on natural conditions These ancient hydraulic works were well designed and built based on careful investigation and analysis on the advantages and disadvantages of natural and economic factors and are successful examples to ancient Chinese people reasonably utilizing and transforming nature to control disasters and develop economy, water conservation and transportation. (1) Grand Canal: Based on minor changes of plain terrain and the flow direction of natural river courses, the Grand Canal used different water sources and diversions, namely Tonghui River section flows east; the north canal section flows south-east; the south canal section flows north; the Lu canal section flows north; the central canal section partly flows north and partly flows south; the Li canal section flows south; and the canal section south of the Yantze River flows both north and south. The different forms of diversion indicate that people paid full attention to natural conditions including lakes, low-lying lands, natural river courses and abandoned river courses and tributary courses. This saved money and labour and facilitated diverting flood water to prevent flooding. The Grand Canal is a good example of taking full advantage of local terrains and water systems. (2) Dujiang Dam: It has its peculiar way of taking advantage of natural terrains. Taking into full consideration the terrain of Minjiang River's outlet, people built the Fish Mouth, internal and external watershed dykes, Sand-flying Dyke and Precious Bottle Neck, dividing Minjiang River into the outer river, which is a natural river, and inner river, which was dug to irrigate farmland. The outer river is designed to divert flood water in flood seasons. The saying "Water channels should be deep while dikes should be low", concluded by ancient working people when they built Dujiang Dam, is still instructive today. (3) Ling Canal: It has more remarkable tourism earthscientific characteristics. In designing the work connecting Li River of the Pearl River water system and Xiang River of the Yangtze River water system, the ancient people made careful studies on the terrain of Mount Taishimiao—the divide between the two rivers, and the different volumes of Xiang River and Li River, and decided to connect these two large water systems by building a dam to block and introduce water from Xiang River to Li River and let 7/10 of water flow into Xiang River and 3/10 water into Li River through the Huazui diversion dam. Behind Huazui were built large and small stone dams to increase load capacity of the diversion dam, and flood water could flow over the stone dams and into the old course of Xiang River.

Outstanding example of utilizing and transforming natural conditions The Grand Canal demonstrated the comprehensive geographical effects of transforming nature and adapting to geographical environments. The Grand Canal was dug in view of different levels of regional economic development and specific transportation and geographical conditions in China. As is well known, after the Sui and Tang Dynasties, areas in the downstream Yangtze River were a major granary of China. However, almost all the dynasties which united the whole country set their capitals in the north, and capitals relied on food supply from the south of Yangtze River. Transporting food from the south to the north was vital to the dynasties in Central Plains, so it was imperative to excavate canals. In terms of transportation geography, main rivers in China all flow from west to east; before marine transport is well developed, south–north traffic had to depend on the Grand Canal and several south–north trunk roads. However, land transport was subject to many limitations including inadequate capacity and was especially unsuitable for fragile goods like porcelains; by comparison, water transport excelled by large capacity and low cost and was more suitable for fragile goods. All these demonstrate the great geographical significance and important role of the Grand Canal in promoting economic and cultural exchanges between the south and the north. Karezes were dug to suit the arid climate of Xinjiang. To store and utilize melted water and reduce water loss by evaporation, people adopted this distinct form of hydraulic architecture. In summer, most of the melted water flowing out of mountains penetrates into the gravel layers of the diluvial fans in front of the mountains and turns into groundwater. Water transfer through open-air channels is impossible because of excessive evaporation and serious leakage, and ordinary wells can hardly gather sufficient groundwater. In view of the special local natural conditions and with reference to the sinking technique of Central Plains, ancient Chinese people originated the method of irrigating with karezes, which utilizes the groundwater rationally.

4.3 Earthscientific Characteristics of Religious Culture Landscapes

Religion is closely related to modern tourism, and pilgrimage itself is a kind of religious tourism program. Therefore, religious culture landscape is a sort of important cultural tourism resource, which has drawn the attention of the tourism circle because of its special and mysterious charm for tourists. The emergence, propagation and development of religion are influenced by social and geographical factors, so it is also of great significance to study it from the perspective of tourism earth-science.

Marxists hold that religion is a kind of social ideology and is part of the superstructure. Religion preaches shadowy and supernatural spirits and is an illusory and reversed reflection of the real world in people's mind. Therefore, the rulers of all dynasties used religion to poison and dupe the people in order to maintain their governance. There are many types and sects of religion in the world. According to the objects of faith, religions can be divided into fetishism, pantheism and monotheism; according to the spheres of influence, they can be classified as primitive religions, Judaism, Hinduism, Taoism and Buddhism, the three worldwide religions of Buddhism, Christianism and Islamism and so on.

On account of political, economic and cultural differences throughout the world, the form and development of religious activities and cultural landscapes present obvious local characteristics and also have lots of tourism earthscientific characteristics which are worthy of study.

4.3.1   Regional Features of Distribution of Religions

Ethnic groups distributed in different regions often have different religious beliefs. However, with the propagation of religious culture and development of modern transportation, cross-distribution of religion has become a trend. Owing to differences in history, politics, customs and other conditions, ethnic groups in different areas have their respective main religions. For instance, Buddhism, which originated in ancient India and Nepal, had Hinayana and Mahayana and many other sects after being spread southwards and northwards. People in the areas along these two propagation routes mainly practise Buddhism (including Lamaism). Therefore, Buddhism today is basically believed in East and South Asia, and the main countries practising Buddhism include India, Nepal, Burma, Thailand, Vietnam, Cambodia, China, North Korea, Japan and so on. Another example is Christianism, which appeared first in Palestine, propagated to Balkan, Eastern Europe, Central Europe and Western Europe in the form of East and West religions (Orthodox Church and Catholicism) after being taken as the state religion by ancient Roman Empire. So the areas where people believe in Christianism are mainly in Europe, including Italy, Spain, the UK, France, Germany and countries in Eastern Europe. In addition, North America, Oceania, South Africa, Latin America and Philippines in Asia have also become important distribution regions of Christianism. As for Islamism, it is mainly concentrated in West Asia, South Asia and North Africa, in which Palestine, Iran, Turkey, Saudi Arabia, Yemen, Egypt, Tunis, Algeria and Morocco are the main Islamic countries.

4.3.2   Regional Features of Religious Activities

Religious activities are those organized by various sects to preach and teach doctrines, canons and scriptures. These activities are actually not tourism resources, and the sects do not take them as the content and form to attract tourists, either. However, those who do not believe in religion have great interest in watching religious activities. In this sense, religious activities constitute a part of cultural tourism resources and we call them pilgrimage tourism resources. So religious activities as tourism resources are targeted at special groups of people. People who believe in religion are opposed to tourism activities based on pilgrimage tourism activities, while those who do not believe in religion are keen on them, which differentiates religious tourism resources from others. Different sects have different regulations for their religious activities. For example, Buddhism has rituals of chanting and Buddhist service, Zen's sitting quietly facing the wall and Tiantai Sect's deep meditation; Hinduism has activities like sprinkling "holy water" and distributing incense ash; Christianism has Christmas and Easter; and Islamism has "Five Pillars" (reciting Islamic creed, worship, fast, Zakat and pilgrimage) and so on. These are very attractive to people who do not believe in religion because of the sense of mystery and novelty.

Influenced by local customs, connotations of religious activities in the world have local characteristics besides common forms of various religions. Therefore, people can meet their different pursuits and pleasures by watching religious activities in different regions. For example, Buddhist and Taoist activities are mostly seen in densely populated towns and picturesque mountain areas, which are rich in tourist sources and in compliance with the doctrine of diving into purifying the heart and keeping away from the madding crowd. Therefore, religious activities are mainly concentrated in famous mountains and deep valleys, as is manifested by the saying "Religion precedes because of mountains; mountain is famous for religion" and the saying "Most famous mountains are occupied by monks". Four famous Buddhist mountains comprising Mount Emei, Mount Wutai, Jiuhua Mountain and Putuo Mountain, and Mount Songshan, Thousand-Buddha Mountain and Jizu Mountain are all tourist attractions famous for Buddhist activities. By contrast, Dragon and Tiger Mountain, Sanqing Mountain, Qiyun Mountain, Qingcheng Mountain and Laoshan Mountain are tour centres known for Taoism.

4.3.3 Regional Features of Religious Cultures

Though religion is a kind of social ideology which worships supernatural spirits, the culture created by it embodies human civilization, reflects the features of national culture and constitutes an important part of the national cultural and historical landscapes of a country. Religious cultures include architecture, sculpture, painting and so on, which are typically represented by three types of architectures in China's Buddhist culture landscapes: temple, grotto and ancient pagoda. Most religious buildings are set up based on local conditions and integrate artificial beauty with natural beauty. Regardless of site selection for churches in famous mountains or site selection, layout, shapes and materials of grottoes, temples, Buddha statues and pagodas, they are all based on ingenuous use of superior natural conditions in order to create a strong religious atmosphere and achieve the goal of long-term preservation. Some buildings are even set up by virtue of profound astronomical and optical wonders. Take for example "Tian Han Bao Yue" (Precious Moonlight) of Caoxi Temple of Anning, Yunnan, "Jin Ding BaoGuang" (Buddha's Light at Golden Peak) of Mount Emei in Sichuan and so on, all of which form mysterious and tempting environments under the religious atmosphere. It can be seen that religious buildings have different features at different ages and in different regions. Ancient people created limitless religious cultures and arts in limited space, which form valuable and excellent heritage of Chinese religious architectures. Temples are set off by mountain scenery. Such a spatial combination is solemn and remarkable. The magnificent temples at different heights give tourists great artistic satisfaction.

The distribution of religious buildings follows the law of earth-science. Take mosques as an example. They are distributed in ghettoes of minority nationalities like Ningxia, Qinghai and Xinjiang in north-west China where people believe in Islamism; and in ancient capitals and coastal cities where there were once many Arabian immigrants, for example Qingjing Mosque in Quanzhou which was built in the Song Dynasty (960AD–1279AD), Qingjiao Mosque in Hangzhou which was founded in the Yuan Dynasty and Jingjue Mosque in Nanjing, Qingjing Mosque in Xi'an and Dongsi Mosque in Beijing. Other examples are Buddhist grottoes and cliffside images, the distribution of which has the following features: (1) the earliest grottoes were mostly built along the Silk Road in ancient times, such as Kizil Thousand-Buddha Caves in Xinjiang, Tura Thousand-Buddha Caves in Kuqa, Mogao Grottoes in Dunhuang and so on; (2) during the Northern and Southern Dynasties, especially Northern Dynasties, most of the large grottoes were sculpted near respective capitals, for example Yungang Grottoes near Datong, capital of Northern Wei Dynasty, Longmen Grottoes in Luoyang, another capital of Northern Wei Dynasty and so on; (3) cliffside inscriptions and images in South China are mostly located near scenic Buddhist temples, such as Leshan Giant Buddha in Sichuan and Thousand Buddha Rock of Qixia Temple in Nanjing; (4) grottoes mostly lean against cliffs and face rivers. Steep cliffs highlight the grandeur of the grottoes, and rolling water sets off a kind of perilous wonder. The mountains and rivers here combine to form Xingsheng. Lithology of grottoes and cliffside images has great influence on the shapes and preservation of Buddha statues. Limestone is dense, so the shapes of limestone Buddha statues are fine and smooth, e.g. Feilaifeng Cliffside Sculptures in Hangzhou and Xiangtangshan Grottoes in Handan. The lithology of Yungang Grottoes in Datong is Jurassic sandstone, which is hard but easy to weather and peel off, so some Buddha statues have already become incomplete with broken legs and noses. Mount Mingsha, where Mogao Grottoes is located, is made up of hard and brittle conglomerates which can hardly remain intact for long, so craftsmen in ancient times collected Achnatherumsplendens, timbers and yellow mud locally to sculpt Buddha statues, which are outstanding examples of making Buddha statues based on local conditions.

4.4 Earthscientific Characteristics of Ethnic Customs

According to historical documents, the habitude formed in a specific natural environment is called "feng", and the convention formed in a specific social environment is called "su". Obviously, as one of the cultural tourism resources, ethnic customs are also a research subject of tourism earth-science.

4.4.1 Analysis of Earthscientific Factors of Ethnic Customs as Tourism Resources

Ethnic customs refer to traditions of a nation in material culture, spiritual culture, family, wedding and funeral and other aspects of social life. Such traditions are habitudes and conventions formed by people of all ethnic groups over a long time under particular natural and historical conditions and are specifically reflected in such aspects such as housing, diet, clothes, wedding, funeral, festival, recreation and sport, and entertaining of guests.

The formation of any ethnic custom is determined by social and historical conditions and regional natural environment. Social and historical conditions refer to the social and historical environments of ethnic groups. The historical development of different ethnic groups is uneven, and the customs formed at different stages of social and historical development are sure to leave their marks. For example, the totem belief of Gaoshan nationality taking snake as their ancestor is the remnant of a belief custom of the primitive society. Before the founding of New China, the wedding custom of "wife not living in her husband's home" preserved by Zhuang, Bouyei, Dong and other nationalities is a remnant of the transition from matrilocality under "matriarchy" to patrilocality under "patriarchy" in the primitive society.

Moreover, important events and historical figures in ethnic history also influenced the formation of ethnic customs. For instance, the "Ongkor Festival" held by Tibetan people before autumn harvest every year is a festival passed down from 1300 years ago when Tibetan people defeated foreign invasion in protecting autumn harvest. The complete set of feudal ethics of Han nationality is ascribable to the profound influences of Confucius and his Confucianism in the spring and autumn period. Mongolians believe in and worship Genghis Khan because he is the hero of the Mongolian people.

The influence of social history on ethnic customs is also manifested in certain social economic base and ethnic culture and art tradition.

The different economic bases of different ethnic groups are bound to nurture different customs. For example, Oroqen and Hezhe people, who mainly engaged in hunting and fishing in history, were clothed mostly in animal and fish skin and lived mainly on animal and fish meat. Some ethnic customs indicate the ethnic cultural art, which was preserved and developed through such ethnic customs. For instance, the people had no characters of their own in the past and they passed down folk songs, a form of oral folk literature, from generation to generation. They not only sang during festivals but also sang instead of speaking in daily life and labour, constantly enriching and developing folk songs.

Natural environment also influences the form of ethnic customs in various aspects. For instance, people who live in Loess Plateau are used to excavating loess caves as dwelling places; people of Dai nationality in Xishuangbanna build stilt-style bamboo houses in order to protect against dampness, poisonous snakes and savage beasts. Local natural conditions usually determine the economic activities of an ethnic group. For example, the ethnic groups who live in semi-arid grasslands rely mainly on animal husbandry and live mostly on beef, mutton and dairy products, while the people living in southern Chinese riverside villages, which are suitable for the growth of rice, live on rice. Natural environment also has profound influence on ethnic culture and art. Take Han nationality for example: owing to differences in the living conditions in the south and north, its cultural and art styles are generally featured by "grace" in the south and "grandeur" in the north.

4.4.2 Earthscientific Characteristics of Civilian Houses

1. Mongolian yurts and tents: Mongolian yurt is a special term used by Man people for the houses of Mongolian nomads. "Yurt" means "home" or "room" in Manchu language. It is like a vault, hence the name Qionglu (literally, sky). Covered by felt, it is also called "felt tent" or "felt yurt", which is "Hana" in Mongolian language. Mongolian yurts are set up both simply and scientifically. Around a Mongolian yurt is a circular framework bound with strips of wood, with a door at one side. The vault-like top is bound with rafters, with a skylight opened in the centre of the top. It is covered with wool felt in winter to keep out cold and with canvas or wicker in summer to let out heat. Meanwhile, it can be easily relocated and therefore is suitable for nomadic life.

   The tents of Tibetan nomads are mostly square, rectangular and oval. The frameworks are built with timber, and the tents are woven with black yak hair. The four sides are pulled by yak ropes and fixed on the ground. The top of the tent is usually double-sloping style with a gap in the centre to let in light and air. It is warm in winter and cool in summer in such tents, and they can be relocated at any time and therefore are quite suitable for grassland life.

2. Stilt-style houses, which are also called "Malan" and commonly known as "bamboo houses", were originally a kind of semi-nested old houses in humid and hot areas in South China. Some ethnic groups in south-west China like Zhuang, Derung, Lisu, Lahu, Wa, Bulang, Dai and Hani are still living in such buildings. Stilt-style houses have two storeys and are erected with timber or bamboo. The upper storey is laid with floorboards and is used as wooden house, and the lower storey has no wallboard around and can be used as stall

or warehouse. Such buildings have four advantages: (1) protecting against dampness and moisture; (2) defending against beasts and poisonous snakes; (3) having good ventilation to drive away summer heat; and (4) being built along valleys, hills and slopes, not restricted by terrain and saving land.

3. Semi-crypt houses were the typical dwelling houses in Yangshao period and are still being used by "Yamei people" of Orchid Island in Taiwan Province and Li people of Wuzhishan in Hainan Island. Orchid Island is close to the ocean and features a hot climate with frequent rains and heavy storms. As above-ground buildings are easily destroyed, Yamei people dig pits under the ground and build houses in them on the hillside by the sea. The walls are built with thick boards, and the outer walls, laid with pebbles, are very firm. The floor inside is paved with pebbles to protect against dampness, and wooden boards are laid on it as beds. The semi-crypt houses of Li nationality are built also to suit the hot and rainy climate and the environment with thick mountain forests. The framework of a stilt-style house is built with bamboo branches, and the arched top is usually covered with thatch. The interior is concave downwards without windows, and the eaves stretch for a long way, so that the house looks like a boat, hence the name "boat-shaped house". Lower eaves can withstand storms and long eaves can provide shade and prevent sunstroke.

4. Adobe arched houses in "Fire Land" Turpan in Xinjiang is the hottest inland basin in China, with more than 100 torrid days in summer and the highest absolute temperature of 47.6 °C, hence the names "Fire Land" and "Hot Pole". To protect against sunstroke, local people in rural areas usually build arched houses with adobe, and some people set basements or semi-basements or dig pits in courtyards to avoid summer heat. Urban residents usually live in flat-roofed houses with earth walls and bearers. The 10–15-cm thick roof are covered with straw on rafters, then plastered with yellow mud and covered with loess and finally smoothed up with mixtures of grass and yellow mud. Such buildings are especially suitable for hot and rainless natural environments.

5. Blockhouses of Tibetan people: Blockhouses are traditional houses built by Tibetans living in Qinghai–Tibet Plateau to adapt to the cold climate. They are usually buildings with elevated bottom floors, the rooms are square and separated by walls, and thick walls and few windows are set to keep out wind and cold; they have earth and wood or stone and wood structures, and the flat roofs can be used for drying grain and protecting against strong wind; and they are usually built leeward and sunward.

6. "Immortal pillars" of Ewenki people in the primeval forests of Great Xing'an Mountain, Ewenki people who earn their livelihood mainly by hunting often lived in houses with conical framework built with more than 30 pieces of thin wood due to unfavourable economic and traffic conditions and in order to suit their hunting life. Such houses are covered with birch barks in summer and animal skins in winter and have a skylight on the top for smoke emission and a fire pit in the interior for sleeping on the floor. Such a house is called "immortal pillar" ("dwelling place providing shelter from sunshine") in Oroqen language, and the Han people call it "Cuo Luo Zi".

7. Quadrangle dwelling houses represented by Beijing Quadrangle are a kind of traditional dwelling unique to the north. It reflects the natural, historical and environmental conditions characterized by the need to protect against cold in winter, sunstroke in summer and wind in spring in economically developed, mid-temperate zones such as Beijing and the areas around it. Such a dwelling usually faces south and has north–south vertical axis. With the principal room in the north, wing rooms at each side and some rooms in the south, it forms an enclosed yard surrounded by houses, hence the name "Quadrangle". As it is tightly structured and has thick back wall, it can keep out thieves and cold.

8. "Yard-style" dwellings south of the Yangtze River: The Yangtze River Valley is located in the subtropical zone, and the dwellings are mostly "yard style". In layout, they are enclosed and also have vertical axis, but they are different from the quadrangles in the north. Not all of such dwellings face south; big yards are usually designed with entrance hall, sedan hall and other rooms on the vertical axis in the middle, and there are living rooms, studies, kitchens, utility rooms and other rooms at the right and left of the vertical axis; areas south of the Yangtze River feature wet weather, and to reduce solar radiation, the yards are designed such that they are wider from west to east than from north to south, and the enclosing walls are tall; and leaking windows are designed for ventilation, and there are also windows in the back walls of the houses.

4.4.3 Earthscientific Characteristics of Other Ethnic Customs

Ethnic customs include clothes, cuisines, festivals and cultural activities as well as residential buildings, all deeply influenced by various regional factors.

1. Locality of clothes: For example, (1) Hezhe people living in the Heilongjiang River Valley, to adapt to their hunting and fishing life and the frigid climate with long winter, mainly wear leather clothing, such as fish skin coats and hats made from skins of fish, roe and deer in winter and plucked leather clothing with buttons on the right in summer. (2) Mongolians living on the grasslands of Inner Mongolian like wearing loose gowns in different colours and with embroidered borders in order to adapt to the grassland climate with long winter and frequent chilly winds, and their horseback economic activities. (3) Tibetans living in areas with high altitude and cold climate often wear special hats known as "xiamaojiasai", meaning "Handi Golden Silk Hat". Such a hat has four brims sewn with fur, with the front and back brims larger than the left and right brims. Men often fold the left, right and back brims inside the hat and only leave the front brim outside, but on snowy days, they leave all the four brims outside. Women usually fold the front and back brims inside the hat and only leave the small left and right brims outside. Elderly people often leave all the four brims outside. Such hats can protect against ultraviolet radiation in

high mountains and keep out wind and snow. These hats are amazingly warm and adaptable. (4) Of the Dai people living in Xishuangbanna, men often wear collarless blouses with buttons down the front or with buttons on the right and small sleeves and wear long trousers, to facilitate ventilation and avoid heat, and women wear white, scarlet and sky-blue tight undergarments, crew-neck shirts with tight sleeves and colourful long skirts.

2. Locality of cuisines: Different regions have their own distinct local cuisines. Such differences result from local economic and production activities, which in turn are restricted by natural environments. Natural environments differ from region to region, leading to different eating habits. For instance, (1) people in the countryside of north-east China live on sorghum and corn meal and like drinking liquor to fight against cold. (2) In areas north of Huaihe River and Qinling Mountain, people generally live on wheaten food. Besides popular foods such as noodle, dumpling, bun, steamed bread and pancake, there are also some other foods with local flavours such as Taiyuan's sliced noodles, Tianjin's Goubuli stuffed bun, Xi'an's shredded pancake in mutton broth and so on; in places like Shanxi and Shaanxi on the loess plateaus, people eat millet and corn (maize) as their staple foods, and other foods include milled glutinous broomcorn millet fried rice cake, Shanxi mature vinegar and soaked milled non-glutinous broomcorn millet in Hetao. (3) In areas south of Huaihe River and Qinling Mountain, the differences in eating habits are even bigger. In Sichuan Basin, which features wet and humid climate, people prefer spicy and hot foods; in the middle and lower reaches of the Yangtze River and the Pearl River Basin, people live on rice and a great variety of non-staple foods. They generally like sweet, sour and hot foods, and the soups are usually light and fresh. (4) With regard to minority nationalities, they have even distinctive eating habits due to special environmental conditions, e.g. dried fish of Hezhe people, mutton eaten with hands of Mongolians, braised meat and horse meat sausage of Kazakh people, double-boiled lamb soup of Tajik people, butter tea, highland barley wine and zanba of Tibetans, sweet glutinous rice of Dai people, scented rice of Li people, fish in vinegar gravy and meat in soy sauce paste of Dong people, sour meat and pickle of Miao people, crusty pancake and braised mutton barbecue of Uygur people.

3. Locality of festival celebrations: All nationalities have their own traditional festive activities with a variety of forms and types, e.g. Spring Festival, Lantern Festival, Tomb-sweeping Day, Dragon Boat Festival, Mid-Autumn Festival and other festivals of Han people, Water-splashing Festival (Dai people), Third Month Fair (Bai people), Song Festival and Frog Respecting Festival (Zhuang people), the Butter Lamp Festival and Lingka Festival (Tibetan people), Festival of Fast Breaking and Feast of the Sacrifice (nationalities believing in Islamism), Duan Festival (Shui people), Munao Festival (Jingpo people), Kuota Festival (Lahu people), Moon Festival (Korean people), Danu Festival (Yao people), and Torch Festival (some minority nationalities in south-west China) (Luo 1986).These festivals all have strong ethnic and local characteristics.

Festival activities of different nationalities are influenced not only by social, economic, cultural, religious and other humane factors, but also by natural factors such as climate, terrain, and hydrology. For example, Han people go for an outing during Tomb-sweeping Day in warm spring, when the flowers are in full bloom and everything come back to life, and they climb mountains on the Double Ninth Festival in autumn, when the weather is agreeable and the atmosphere is highly transparent. Miao people living in the mountainous areas of south-west China have "Stepping on the Colourful Mountain" Festival. Mongolians have the "Nadam Fair" including wrestling because the green grasslands provide ideal places for wrestling and horseback archery for Mongolian nomads who live on Inner Mongolian Grassland.

References

Kuang, G. (1987). *An introduction to human geography*. Chongqing: Southwest China Normal University Press.

Luo, Z. (1985). *Chinese ancient pagodas* (p. 21). Beijing: China Youth Press.

Luo, J. et al. (1986). *Chinese traditional festivals*. Beijing: Beijing Popular Science Press.

Ma, Z. et al. (1987). *Brief introduction to Chinese historical geography* (pp. 336–339). Xian: Shaanxi People's Publishing House.

Nianhai, S. (1981). *Historical changes of beautiful landscapes of our motherland, collected theses on Chinese historical geography* (p. 39). Beijing: Shaanxi People's Publishing House.

Yu, X. et al. (1985). *Land and people of Beijing*. Beijing: China Travel and Tourism Press.

Zhengxiang, C. (1983). *Chinese cultural geography* (p. 1). Beijing: SDX Joint Publishing Company.

Chapter 5
Earthscientific Characteristics
of Tourism Resources

5.1 Research Value of Earthscientific Characteristics
of Tourism Resources

It has been proved that the recognition, exploitation and utilization of tourism resources are inseparable from earth-science. However, at present relevant literature seldom discusses characteristics of tourism resources exclusively from the earthscientific perspective, but just touches upon earth-science in the discussion on tourism or features of tourism resources. When discussing modern tourism, Guo Laixi puts forward such characteristics as popularity, economy, integrity, diversity, locality and seasonality[1]; Lu Yunting points out that tourism resources have eight features (Yunting 1988): universality, modernity, variability, locality, intersection, sustainability, seasonality and the function of promoting intellect; Liu Zhenli holds that the characteristics of tourism resources are locality and integrity (Zhenli 1987); and Lei Mingde argues that the characteristics of tourism resources are universality, locality, natural occurrence and historical survival, modernity, seasonality, culture, dynamics and scientificity (Mingde 1988). It is not difficult to see that some features mentioned above, e.g. locality and seasonality, are the earthscientific characteristics of tourism resources.

This book studies the characteristics of tourism resources exclusively from the earthscientific perspective, so as to further illustrate earthscientific theories and methods; help tourism developers understand the earthscientific characteristics and laws of tourism resources and apply them to the recognition, discovery, development, utilization, protection, improvement and creation of tourism resources and the development of tourism; and help tourists apply these laws in tourism acts and activities to enhance aesthetic and appreciation abilities, arouse interest in tourism,

---

[1] Tourism Geography Group of Institute of Geography, CAS, 1982, The Proceedings of Tourism Geography, 23.

© Springer-Verlag Berlin Heidelberg and Science Press Ltd. 2015
A. Chen et al., *The Principles of Geotourism*, Springer Geography,
DOI 10.1007/978-3-662-46697-1_5

strengthen tourism awareness and enter a deeper and higher level of spiritual and material enjoyment, so that tourist activities will make people more confident about their future life and work and contribute more to social progress.

The generation, distribution, evolution and development of tourism resources all follow the basic laws of earth-science. Therefore, only if people understand earthscientific theories and methods, can they apply them to the investigation and evaluation of tourism resources, formulation of tourism plans, division of tourism areas, selection of tourism routes, full use of tourist seasons and even the protection and re-creation of tourism resources, etc.

5.2 General Earthscientific Characteristics of Tourism Resources

Earthscientific characteristics of tourism resources can be summarized as the following six categories: regional differentiation characteristic, scenic combination characteristic, similar appearance characteristic, potential resource characteristic, seasonal variation characteristic and resource nature variation characteristic.

5.2.1 Regional Differentiation Characteristic

Regional differentiation characteristic is the most fundamental earthscientific characteristic of tourism resources and the most fundamental factor resulting in tourism resources. If all the areas of the world are the same, there will be no tourist motivation; if a tourist site does not have its unique scenery, it cannot attract many tourists. In discussing tourism characteristics, Yu Guangyuan fist mentioned allopatry, which is the tourism nature determined by regional differentiation of tourism resources. Regional differentiation of tourism resources is controlled by natural geography and the law of social activities, with the former playing a more direct and obvious role.

5.2.1.1 Tourism Resources Have the Distinct Distribution Characteristic of Latitude Zonality

Latitude zonality of tourism resources is mainly caused by climatic differences. From the equator to the poles, there are tropics, subtropics, temperate zones and frigid zones with great changes in temperature and heat.

In the tropics, there are luxuriant vegetation, evergreen trees and a large number and variety of animals. In tropical marines, there are also a great variety of living things and picturesque submarine scenery. In tropics and subtropics featuring high temperature, abundant rainfall and warm and wet climate, karst is also well developed on the surface and under the ground. Therefore, on the ground, there are large areas of peak forest landforms including solitary peak, stone forest and stone bud; under the ground, there are well-developed karst caves, underground rivers, stalagmites, stone columns, travertine and stalactites, e.g. Lunan Stone Forest in Yunnan and Guilin scenery in Guangxi. Owing to agreeable sea temperature and adequate sunshine, coastal areas located at medium and low latitudes are excellent bathing beaches and good destinations for various maritime tourist activities. As mentioned above, currently the most attractive tourist activity is visiting "3S (spring, sport, shopping)" landscapes, which are mainly distributed in Mediterranean Sea, Bay of Bengal, South China Sea, Hawaiian Islands, Caribbean Sea and other coastal areas at medium and low latitudes. Spanish people say, "What we export are sunshine, air and beaches". Hainan of China is also an ideal tourism site with "3S" landscapes, beautiful natural scenery and special ethnic folklore. Submarine sightseeing in the tropics is also very popular. For instance, more than 30 countries including the United States, Australia, Singapore, Thailand, the Philippines and Indonesia have established more than 280 submarine tourist centres, which organize submarine sightseeing, underwater hunting, photographing and salvage and other activities. China has also set about developing submarine tourism programmes in Zhanjiang, Dianbai, Hainan Island and other places and has made preliminary achievements.

In the temperate zones, owing to low temperature in winter, the vegetation features deciduous broad-leaf forest and mixed coniferous and broad-leaf forest, the number and variety of animals, plants and marine lives are much smaller, but in some places, there are a large number or special types of fish and shrimp resources. Owing to low temperature, inadequate rain and lower precipitation intensity, karst is not developed aboveground but is well developed underground. For example, there are quite good underground karst landforms in North China, Causses Plateau in France, Czechoslovakia and other places. Moreover, the opening time of bathing beach is influenced by short summer in temperate zones. However, in the northern hemisphere, as the mid-latitude areas have the largest land area, most mountains and densest population distribution, they are very rich in tourism resources and also have many famous mountains, rivers and cultural relics, e.g. Mount Fuji and Oriental Land in Japan, Yellow Stone National Park, Grand Canyon Colorado and Disneyland in the United States, Bavarian Park in West Germany, pyramids in Egypt, Ancient Athens in Greece, Site of Ancient Rome in Italy, Versailles Palace in France, Buckingham Palace in England, many famous mountains and ancient cities in China and famous historic sites in the world.

In frigid zones and chilly polar regions, there are coniferous forests and ice sheets. The number and variety of living things here are small, but unique animals such as seal, walrus and penguin, and ice and snow and polar lights are all extremely precious tourism resources. People generally go there for energy-intensive sports such as skating, skiing, hunting and exploration and for viewing polar lights, polar day and polar night. The Snow Festival in Hokkaido, Japan, the ice sculpture in Heilongjiang, China, and the polar hunting and polar lights sightseeing in Finland and Sweden are all particularly attractive.

5.2.1.2 Tourism Resources Have the Distinct Distribution Characteristic of Longitude Zonality

Distribution of longitude zonality is mainly restricted by the aridity and humidity of climate. Generally speaking, the farther the place is away from the ocean, the less the atmospheric precipitation, the poorer the vegetation coverage, the drier the climate and the bigger the wind–sand action will be. Tourism resources of the sea-side are very attractive to people living inland; therefore, many people move from inland area to the seaside for travel. Off the shore are usually plains, rivers and lakes, where most large- and medium-sized cities are political, cultural, economic and transportation centres, and are therefore rich in cultural tourism resources. Further inland, natural landscapes gradually change from forests to plains, here used to be political and cultural centres in ancient times and so there are plenty of historical and cultural tourism resources such as temples, palaces, monuments and pagodas, stone carvings and murals. Still inward towards the inland centre, the climate is very dry and the natural landscapes change from grasslands to deserts. Here wind–sand action results in residual hills and strange stones, forming wind-erosion mushrooms, wind-erosion columns, wind-erosion castles and so on, e.g. Urho Windy City in Xinjiang and General Gobi in the north of Qitai, etc. Most ancient cultures in inland areas have been buried by wind and sand; therefore, the cultural relics gradually unearthed and the historical sites exposed on the ground are great attractions for tourists. Inland tourism nowadays provides a huge space for high-level culture tourists who take pains to pursue the development of human culture, e.g. the "Silk Road" tourism, Great Wall tourism and sightseeing of historic sites of Qin and Han Dynasties of China.

5.2.1.3 Tourism Resources Have the Distinct Distribution Characteristic of Vertical Zonality

Such a characteristic is due to temperature and heat changing with height. The band spectrum of natural landscapes is the same as that from the tropics to frigid areas. Therefore, Mount Everest, the highest mountain on the earth, is called the third pole in the world. Due to sharp uplift of mountains in the transition zones between plains and mountain regions, perilous and miraculous precipices, cliffs, canyons and abysses are formed, which usually become scenic spots with forests, secluded torrents, waterfalls, deep pools, hot springs and streams. These scenic wonders were frequented by emperors and nobles in ancient times and are where Chinese Buddhist and Taoist temples are located and the men of letters of all times travelled and expressed feelings and inspirations. Many famous mountains in China are distributed in these zones. In sparsely populated, less damaged medium and high mountainous regions, many places remain in their virgin state and are ideal choices for establishing national nature reserves and forest parks. In these places are many deep valleys and perilous dangerous shoals, which are suitable for venture lovers to enjoy drifting and expeditions. In high mountainous regions that are mostly covered by tundra or glaciers, the climate is very cold, the air is thin, and few people have gone there since ancient times. However, they are ideal destinations for mountaineering tours, scientific expeditions and ice and snow sports. Every year, Mount Fuji, Alps, Tashkent, Tianshan, Pamir, Himalayas, etc. attract swarms of mountaineers.

5.2.1.4   Tourism Resources Have the Distinct Distribution Characteristic of Centrality

This characteristic results from the combined action of the earth's internal and external factors. Tourism resources may concentrate in large areas and small areas or gather in blocks and bands.

As mentioned above, karst landscapes only appear in areas with carbonate formations. Blocky distribution of carbonate formations on the earth leads to blocky distribution of karst landscapes. Karst landscapes in China are mainly distributed in Guangxi, Guizhou, Yunnan, Hunan, Western Hubei, Eastern Sichuan, Western Qinling, Mount Taihang and border regions between Zhejiang, Anhui and Jiangsu.

Volcanoes are distributed like bands. In New Zealand, there are many unique volcanic parks, of which Tongariro Volcanic Park has 15 craters. Near San Jose, capital of Costa Rica, there are hundreds of volcano tourist areas, and the Perth Volcano is the largest geyser volcano in the world, with boiling water shooting to a height of more than 800 m. In the provinces of Jilin and Heilongjiang, China, there are more than 230 volcanoes, over 100 of which are located in Changbai Mountains. Mineral springs also have the distinct characteristic of concentrated distribution. In China, there are more than 2600 mineral springs, of which over 630 are in Tibet, over 440 in Yunnan and over 500 in Guangdong, Fujian and Taiwan. The distribution of volcanoes and springs is closely related to faulted structure and their fractures.

China is a country with lots of lakes, which are mainly distributed in Qinghai–Tibet Plateau and eastern plains. The distribution relates to both large drainage systems and faulted structures.

Cultural tourism resources are also distributed in blocks and bands. Han culture of China concentrates in central and eastern regions; Tibetan culture concentrates in Qinghai–Tibet Plateau, and cultures of other minority nationalities are also concentrated in remote areas. Along the Great Wall, "Silk Road" and Beijing-Hangzhou Grand Canal, there are related historical cultures. In respect of gardening, there are gardens in North China, gardens south of the Yangtze River, gardens in Lingnan and gardens of minority nationalities, which feature concentrated distribution by different styles.

5.2.2 Scenic Combination Characteristic

An isolated scene is often difficult to be developed and utilized as tourism resource, especially when tourism has developed into a profit-seeking industry. It is well known that the world's biggest iron lion 20 km east of Cangzhou, Hebei, is of great academic significance and should have become an important tourism resource. However, it is seldom-visited for its location in the wilderness. Mount Huashan, because of "only one road leading to Mount Hua", connects five peaks comprising East Peak, South Peak, West Peak, North Peak and Central Peak and many perilous roads and surpassing sceneries along the way like Eighteen Bends, the Thousand-Foot Precipice, Hundred-Foot Crevice, Ear Touching Cliff, Blue Dragon Ridge and South Heaven Gate. All these scenes together with green pines and old trees and such cultural landscapes as Taoist temples and cliffside inscriptions constitute a typical, massive, compact and hierarchical mountain resort. The numerous scenes combine to set off "peril" vividly and thoroughly. For example, the "Ear Touching Cliff", which is just a slightly slanting rock wall and is by no means a good scene, gives people a sense of peril for its setting against the dangerous road on Mount Huashan. Therefore, scenic combination is of great significance.

There are various scenic combinations of tourism resources, e.g. homogeneous combination and comprehensive combination; group combination and trans-regional combination; and large-scale and small-scale combinations.

5.2.2.1 Homogeneous Combinations and Comprehensive Combinations

The combination of tourism resources of the same nature is called homogeneous combination. For instance, the 72 peaks represented by Tiandu Peak, Lotus Flower Peak, Beginning-to-Believe Peak, Bright Summit Peak and Peach Peak (Flying Stone) form the framework of Mount Huangshan, hence the name "72 peaks of Mount Huangshan". The combination of tourism resources with different natures is called comprehensive combination. For example, fantastic pines, grotesque rocks, seas of clouds and hot springs constitute "four wonders" of Mount Huangshan. Another example is that the main part of West Lake is the combination of five lakes comprising Outer Lake, Inner Lake, Yue Lake, West Inner Lake and Small South Lake; around West Lake are more than 40 places of interest and 30 important cultural relics and historic sites. The combination of natural and cultural tourism resources is achieved by the "ten views of West Lake", namely Spring Dawn on the Su Causeway, Autumn Moon over the Calm Lake, Viewing Fish at Flower Pond, Orioles Singing in the Willows, Twin Peaks Piercing the Clouds, Three Pools Mirroring the Moon, Leifeng Pagoda in Evening Glow, Evening Bell Ringing at the Nanping Hill, Winery Yard and Lotus Pool and Lingering Snow on the Broken Bridge.

5.2.2.2 Group Combination and Trans-Regional Combination

There are group combinations of local, nearby or distant tourism resources. Examples of local group combinations are the so-called "Three Three, Six Six" Wuyi scenic wonder composed of Jiuquxi River and 36 peaks and the "Celestial City and Buddhist State" comprising nine peaks including Tiantai Peak, Lianhua Peak, Tianzhu Peak and Shiwang Peak and more than 300 temples such as Huacheng Temple, Incarnation Hall, Longevity Palace, Zhiyuan Temple, Minyuan and Tiantai Temple of Jiuhua Mountain. Examples of nearby combinations are "four wonders in Daiding of Mount Taishan" comprising sun rises from the east, sunset clouds and evening glow, Yellow River Golden Belt, and Sea of Clouds and Jade Plate, and the Himalayan Mountains where there are many mountains higher than 8000 m.

China's five famous mountains, namely Mount Taishan, Mount Huashan, Mount Hengshan (in the north), Mount Songshan and Mount Hengshan (in the south), are distributed in Shandong, Shaanxi, Shanxi, Henan and Hunan. These mountains extend 660 km from east to west and 1400 km from north to south. The Yangtze River and the Yellow River separate North China Plain from Jianghuai Plain and Taihang Mountains from Funiu Mountains. This is a typical trans-regional combination of natural tourism resources. China's four Buddhist holy mountains, namely Mount Wutai, Mount Emei, Jiuhua Mountain and Putuo Mountain located, respectively, in Shanxi, Sichuan, Anhui and Zhejiang constitute a typical trans-regional combination of cultural tourism resources. The "seven wonders" of the world is a trans-regional combination covering the longest distance. Despite the long distance, trans-regional combination of tourism resources enjoys the same reputation, so that people often associate one of them with all others and hope to travel farther.

5.2.2.3 Large-Scale and Small-Scale Combinations

The combination scales of tourism resources differ significantly. Examples of small-scale combinations are the Twin Pagoda Temple (the symbol of Taiyuan) composed of a temple of the Ming Dynasty and two adjoining pagodas, Three Pagodas (in Dali, Yunnan) consisting of one square pagoda and two octagonal white pagodas, and the Pagoda Forest in Shaolin Temple in Mount Songshan. Several groups of granitic landforms discovered by the author recently in Laoling tourism area in Qinhuangdao are also a small-scale combination. Within 500 m of the said combination, there are various characters (Fig. 5.1) in *Journey to the West*, including Tang Priest expecting apprentices, Bajie looking for big brother, Wukong teasing Bajie, White Horse and Ox King.

Typical large-scale combinations include Three Gorges (over 190 km long and composed of Qutang Gorge, Xiling Gorge and Wuxia Gorge together with White Emperor City, Goddess Peak, Gaotangguan Temple, Zhaojun Village and Quyuan's House along the way), "Guilin scenery" (composed of various karst

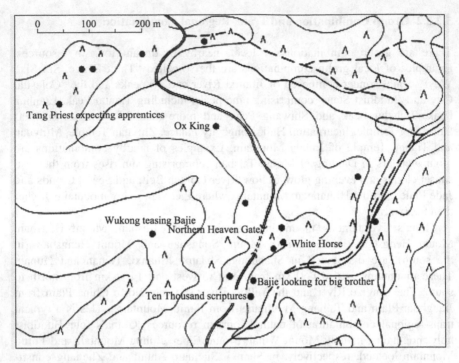

Fig. 5.1 Characters in journey to the West of combined imaginative landforms in Laoling tourism area in Qinhuangdao

landforms within the range of more than 100 km extending from Xing'an in the north to Yangshuo in the south of Guangxi) and the Great Lakes on the frontiers of the United States and Canada.

Tourism cities such as Beijing, Chengde, Qinhuangdao, Xi'an and Nanjing can be regarded as middle-scale combinations of tourism resources.

5.2.3 Similar Appearance Characteristic

Similar natural tourism resources often appear in places wide apart. For example, limestone caves are widely distributed in China, Yugoslavia, Switzerland, France, Czech and many other countries, and the same is true of volcanoes, waterfalls, canyons, beaches and hot springs.

Similar appearance of cultural tourism resources is more common. For example, Kwan Tai Temple, Dragon King Temple, Confucius Temple, Buddhist and Taoist temples can be found all over China; Great Buddha's Hall, Eighteen Disciples of the Buddha and Four Heavenly Guardians can be found almost everywhere. This is even more true of Catholic Churches in Western countries.

Similar appearance of tourism resources enables tourists with limited money and time to visit the nearest similar scenes. However, it also causes problems to the development of tourism. For instance, if foreign tourists in China see Eighteen Disciples of the Buddha in Beijing, Suzhou and Hangzhou, they may be fed up with seeing temples during the day and sleeping at night. For another example, when visiting royal memorial parks, tourists are generally in high spirits when touring the first mausoleum, but they may find that the second one is much the same as the first and will not see the third at all. Obviously, it would be better not to revisit similar scenic spots and it is necessary to step up advertising respective features in order to arouse new desires of tourists.

5.2.4 Potential Resource Characteristic

Man's understanding of the objective world is built up gradually. Whether a landscape with objective tourism value can become a tourism resource is determined by many factors such as the level of understanding, needs of senses, time of discovery, exploration capability, and publicity and opening conditions. A resource that has not been recognized as tourism resource but, as proven in history, is sure to become a tourism resource in the future can only be taken as "potential tourism resource". Mount Taishan and Mount Huashan have a long history as tourist attractions, while Zhangjiajie and Nine-village Valley have been recognized and developed as tourism resources for only several years. Some historical relics appreciated by a small number of people in the past are being opened to the broad masses and displaying bigger potential effect. In recent years, karst caves in China have been discovered and developed into tourism resources successively. After being buried for 2000 years, Emperor Qin's Terra-Cotta Warriors and Horses have become an important tourism resource attracting swarms of visitors from all over the world; the excavation, repair and opening of the Mutianyu Great Wall, Jinshanling Great Wall, Simatai Great Wall, Xifengkou Great Wall, Jiumenkou Great Wall and Laolongtou Great Wall between Beijing and Qinhuangdao are arousing the interest of tourists. In 1987, a group of young French tourists aroused attention by running along the section of the Great Wall between Badaling in Beijing and Laolongtou in Qinhuangdao. It can be predicted that there will appear a hiking fever for Beijing—Qinhuangdao Great Wall that has a comfortable length, enjoys convenient supply and integrates the essence of the Great Wall.

What the local people take for granted may well be very attractive tourism resources for strangers. Therefore, to discover and identify tourism resources is the most important task for tourism developers. The Changli Golden Coast in Hebei has long been an objective existence but remained ignored until Guo Laixi and others found a place with great tourism value, which was later to be named "Golden Coast".

5.2.5 Seasonal Variation Characteristic

The seasonal variation characteristic of tourism resources results in high and low seasons of tourism and brings prosperity and depression to the market and tension and leisure to employees.

Seasonal variations of tourism resources are mainly caused by seasonal climatic variations, especially in middle latitudes. Seasonal changes are often described as bright spring, verdant summer, golden autumn and freezing winter. In fact, people are reluctant to travel in autumn, when the falling leaves give people a sense of sadness despite their golden colour, and in winter, when the chill outside discourages people. Tourists generally like to go out for travel when the temperature remains above 20 °C, which is agreeable for human body. Chinese people have long been in the habit of going out in spring when the earth wears her coat of green, flowers are blooming, and natural landscapes are full of vitality. When the temperature is or above 30 °C, the weather gets hot, but it is relatively cool on seashores or mountainous areas, and it is good time for swimming or water sports in rivers, lakes and seas. Therefore, tourists flock to the seashores and high mountains to avoid summer heat. In May and June, Southeast Asia is in the rainy season and is unbearably sultry, while China is in bright and beautiful spring, so people swarm from Singapore, Thailand and other areas to China. In July and August when Central and North China is extremely hot, seaside cities such as Qinhuangdao, Dalian and Qingdao and summer resorts including Mount Lushan, Mount Huangshan and Mount Emei welcome their tourism peaks.

Winter in middle latitudes is generally a low season for tourism. For example, in the winter of North and North-west China, it seldom snows, the trees wilt, the climate is cold, dry and windy, and the whole natural environment is in "dormant" state. But in areas where it snows much in winter, winter is often a high season for tourism. For example, in Soviet Union, North-east China and Hokkaido of Japan, the nature takes on silvery white in winter and people may conduct such special activities as skating, skiing and appreciating rime (ice feathers) and ice sculptures. As people take greater interest in sports, winter tourism will flourish gradually.

Generally, cultural tourism resources themselves do not change with season, but tourists' moods are usually affected by climatic change. Moreover, some cultural tourism resources only occur in specific periods, e.g. the Spring Festival, Lantern Festival, Dragon Boat Festival, Corban Festival, Water-splashing Festival, Third Month Fair and Nadam Fair in China and Christmas Day and April Fool's Day in Western countries. Other examples are Guangzhou Trade Fair, Weifang International Kite Festival, Harbin Ice Lantern Show and Wuqiao International Acrobatic Art Festival (Hebei), which have become popular in recent years. Tourism developers should develop various cultural tourism resources with local characteristics according to local features and habits so as to prolong the high season and adjust the low season for tourism.

5.2.6 Resource Nature Variation Characteristic

Variability is a dynamic characteristic of tourism resources. Tourism resource is often a form of existence at a particular stage of the development of something. For example, rocks on a hill may be considered as tourism resource only when their shapes have evolved such as they have some visual value (e.g. when they look like a beauty, a mushroom). If a part of the top of the mushroom collapses so that the mushroom becomes a "crutch" or a "boot", it is still a kind of tourism resource despite the change in form and nature. If it collapses into a heap of stone one day, it will lose the nature as tourism resource. But if it is carved into a beauty or a fantastic animal, it will become a tourism resource again. Though Tangshan earthquake in 1976 was a massive tragedy, some remains preserved have become valuable tourism resources. The Imperial Palace in Beijing, which was originally the royal forbidden area for the emperors of the Ming and Qing Dynasties, has become a tourist spot today. Jinggang Mountains and Yan'an where there were flames of war have also become sacred places that the tourists yearn for. Some traditional activities of local nationalities have become tourism resources, attracting a great many tourists from other places. Some former tourist attractions have stopped all tourist activities in order to protect water resources.

In a word, the nature of many tourism resources is changeable. Some non-tourism resources may turn into tourism resources; some tourism resources may lose tourism value and become non-tourism resources; and one kind of tourism resource may change into another kind. Such variations may be utilized but may also ruin the tourism sector.

The first four aspects discussed in this section may also be considered as the characteristics of tourism resources in spatial distribution, the fifth and sixth aspects are the characteristics of tourism resources in time distribution, and the fourth, fifth and sixth aspects may also be regarded as dynamic characteristics of tourism resources.

5.3 Independent Earthscientific Characteristics of Natural Tourism Resources

Besides common characteristics, natural and cultural tourism resources have their respective individual characteristics. Firstly, we discuss the individual characteristics (independent earthscientific characteristics) of natural tourism resources.

5.3.1 Immovability

As the subjects of natural tourism resources, water, mountain, forest and grass are formed under certain geological and geographical conditions. Their morphological characteristics, ecological environments and tourist functions are all unique, non-duplicable and immovable, for example Mount Huashan, which is known for its "peril", Mount Taishan, which is famous for its "majesty", Mount Huangshan and the Three Gorges. All of them have unique characteristics and are irreplaceable and immovable. The same is true of Grand Canyon Colorado in the United States, Victoria Falls in Africa and the Costa del Sol and bright sunlight in Spain. The immovability characteristic of main natural tourism resources guarantees their absolute monopoly in tourism.

The layout of natural tourism resources of an area also forms naturally, and people have to comply with this natural pattern, connect it with reasonable roads and supplement it with necessary man-made sceneries and service facilities to set off the environments. For example, Mount Huashan, Mount Huangshan and Mount Taishan are all tourist attractions built with successful integration into the natural pattern of natural sceneries.

The immovability characteristic of natural tourism resources is not absolute. Migratory birds and fishes breed and rest in different places following fixed routes all the year round, becoming common tourism resources for different regions, but their migratory routes are hard to change; therefore, such tourism resources have an more advanced immovability characteristic.

Mineral water, corallite, rain flower pebble and numerous trees and flowers can also be moved to other places and treated as movable tourism commodity. However, they will promote the tourism of a region instead of significantly affecting it. For example, mineral water of Laoshan Mountain, rain flower pebble in Nanjing and Jatropha multifida in Taiwan all contribute greatly to the enhancement of the popularity of Qingdao, Nanjing and Taiwan.

5.3.2 Periodical Variation Characteristic

The daily variation of some tourism resources is remarkable. Tourism resources such as sunrise in the morning, sunset in the evening, tide (appearing twice in a day), chirping of birds and haunting of beasts all follow the law of daily variation to some extent.

In summer, the air is fresh, and flowers and trees are moist and straight in the morning; it is hot at noon and flowers and trees wilt and droop; and a gentle breeze brings coolness at night. But in another sense, the exquisite scenery in the morning and the sharp contrast of light and shade in the evening make it easy to obtain satisfactory photographic and video effects; though the noon is tedious, it is a great time for swimming in the sea and sunbathing. All these are typical periodical variations of natural tourism resources. Seasonal variations mentioned above are also periodic variations.

5.3.3 Random Variation

Many natural tourism landscapes appear "incidentally" or very rarely, giving people a sense of "eagerness" and mystery. This is the effect of random variation of tourism resources.

It is hardly possible to appreciate the wonderful view of rising sun on the top of Mount Taishan in summer because of cloud and mist in seven out of ten days. It is better to appreciate it from October every year to May of the following year.

Sea of clouds is also rare on Mount Huangshan in summer, because the sea of clouds on Mount Huangshan appears once every 40 days on average, and it appears mostly from November every year to May of the following year and appears least in July and August.

Mount Emei, Mount Lushan, Mount Taishan and Mount Huangshan in China are all famous spots for appreciating Buddha light. Danya Mountain in Penglai is a place where mirage appears, but nobody can predict its appearance. The wonderful view "Shining at Jinhua Mount" in Yesanpo scenic area of Laishui, Hebei, is magnificent with the whole mountain in red. It is said that such a sight once fooled fire fighters in western outskirts of Beijing, who drove many fire engines to "put out the fire" and that it occurs once every 60 years.

5.4 Independent Earthscientific Characteristics of Cultural Tourism Resources

5.4.1 Movability

Cultural tourism resources are created by man and so are generally duplicable and movable. A pavilion, a hall, a castle, etc. are all duplicable or can be moved to another place. The Xi'an Forest of Stone Tablets is a collection of famous tablets from all over the country. The big bell (Huayan Bell) in the Great Bell Temple was originally set in the Printing House of Chinese Buddhist Scriptures in Deshengmen, but was later moved to Wanshou Temple in the western outskirts of Beijing and finally to the present address. Taoranting Park once moved Atelier and Qin Room of Emperor Qianlong from Zhongnanhai. Cultural exchanges, economic transactions and even wars and plunders in the world lead to long-distance displacement of tourism resources. Nowadays, the displacement of cultural tourism resources is amazing both in speed and in scale. The United States built "Disneyland" in Los Angeles in 1955 and opened "Disney World" in Orlando, Florida, in 1971. The "Oriental Land" built by Japan in 1979 is more advanced than "Disneyland" and "Disney World" and can receive 10 million tourists every year. In recent years, paradises of the nature of Disneyland have been coming forth all over the world. Various modern recreational facilities are also being built in many big cities of China. China's garden art, which is recognized in the West as "the Origin of Gardens in the World", has been introduced to Japan, Europe and America for cultural exchange long ago. In recent years, the United States spared no money in duplicating Dianchunyi in Master-Of-Net Garden of Suzhou for Metropolitan Museum of Art, New York. Chinese Garden, one of Singapore's major tourist centres, was built in complete imitation of Chinese gardens. Moreover, those unearthed relics, artistic handicrafts, clothes and delicious dishes have been presented to each other, exhibited successively, and publicized and duplicated for promotion, which contributes greatly to the growth of tourism revenue.

Of course, to protect the dignity of a country or a nation, some influential national treasures and rare cultural relics are not permitted to be transferred or duplicated.

5.4.2 Antiquity

The value of many cultural tourism resources lies in their long history. Tourists visit a country with an ancient civilization primarily because it has a time-honoured culture. "The Great Wall" is reputed as one of the ancient seven wonders of the world for the fact that it reflects Chinese culture over the past 2000 years. Emperor Qin's Terra-Cotta Warriors and Horses is regarded as "the world's largest underground military museum" because it reflects the sculptural arts 2000 years ago. The Colosseum of Rome, Italy, is reputed as one of "the seven wonders of the world" because it reflects the great achievements of architectural art 1900 years ago. Generally speaking, the longer the history of cultural tourism resources is, the higher the tourism value will be.

The tourism value of natural tourism resources is generally not determined by the degree of antiquity. Only when the length of life of a tourism resource is comparable with the history of human culture, can it show some antiquity, e.g. the few locust trees preserved from the Zhou and Tang Dynasties in China.

5.4.3 Spiritual Culture Characteristic

Although natural tourism resources themselves are purely material, spiritual cultures such as different social systems, histories, languages and ethnic customs are an important part of cultural tourism resources. Religious tour is an important tourist flow; Qufu of Shandong, birthplace of Confucianism Culture, attracts a great number of tourists every year; and the legend Lady Meng Jiang's Wailing at the Great Wall adds charm to tourism in Shanhaiguan Great Wall, with Jiangnv Temple crowded with people in high season for tourism. Many tourists travelling in Water Margin Liangshan are inspired by the 108 Liangshan heroes in *Water Margin*, such as Wu Song, Li Kui, Lin Chong and Song Jiang; although tourists know that "Grand View Garden" in Beijing and "Ning and Rong Mansions" in Shijiazhuang are feigned buildings, they still pretend to be Baoyu and Daiyu to take photographs on the spot due to influence by the novel *Dream of the Red Chamber*, especially the namesake TV play series; Grapes Festival in Vevey, Switzerland, is ceremoniously held 4–5 times every 100 years, each time lasting a fortnight; and the Philippines' "Peaceful Reunion Tourism Plan" invites soldiers serving in the army in the Philippines during World War II to return to the Philippines together with their families. Other tourism activities themed on spiritual culture are tourism activities carried out in Hawaii in commemoration of Pearl Harbour Incident and the old battlefields, the dragon boat race in commemoration of the ancient poet Qu Yuan, and the travel activities in the Year of the Dragon in China.

5.4.4 Continuity

Cultural tourism activities often feature historical repeatability or continuity. For example, religious activities, which were originally spontaneous activities of followers of a religion, have become a minor social trend with the development of tourism. Temples that were once desolate have become extraordinarily bustling with kneelers in flocks, with the number of pilgrims ever increasing. Many non-believers involuntarily worship on bended knees in such a mysterious atmosphere. Archaized buildings are built with the tide of fashion; for example, Hong Kong has built "Song Dynasty Town", which makes a pile of money every day. Many places have set the street of Ming Dynasty or the street of Qing Dynasty, built underground palace museums and repaired ancient buildings which have long been deserted, e.g. Yellow Crane Tower in Wuchang, Laolongtou in Qinhuangdao, Pavilion of Prince Teng in Nanchang, Mutianyu Great Wall in Beijing and Jinshanling Great Wall in Luanping, reproducing and preserving ancient cultures that are falling into oblivion. Participatory archaized tourism activities have become the most popular, so royal hunting grounds are being repaired and opened, cuisines in the Palaces of the Qing Dynasty and Confucius Mansion have taken on a new lease of life, and ethnic festivals, customs and activities on ancient battlegrounds are being gradually applied to tourism. In recent years, novels, films and videos describing the Qing Dynasty keep coming out, people get more and more familiar with the history of the last feudal dynasty, and a travel boom is emerging in the "Qing Dynasty" archaized tourism itinerary centred on numerous royal palaces and famous gardens in Beijing and covering Eastern and Western Royal Tombs of the Qing Dynasty, Chengde Imperial Summer Resort, Mulan Paddock, temporary imperial palaces in Jixian and Shanhaiguan and even the Old Palace in Shenyang.

5.4.5 Times Characteristic

Besides existing ancient cultural tourism resources, it is necessary to reform and create new tourism resources conforming to the trend of the times, such as special projects like Disneyland, car and motorcycle rallies, conference tourism, and culture and art tourism, and new projects like tourism fast food, mobile houses for tourism and multi-function touring ships. All of these follow the trend of the times. Some archaized tourism activities also have strong times characteristic, e.g. the international "Silk Road" car rally held in autumn 1985. It connects the ancient "Silk Road" with modern transportation roads and means, enabling tourists to appreciate the natural scenery of North-west China, visit and recollect ancient culture while doing physical exercises and achieving new cultural exchanges. Therefore, this is a typical example of tourism making the past serve the present. The "Silk Road" tour will promote new prosperity there and inspire us to develop more cultural tourism resources and products conforming to the trend of the times.

As modern tourism has become a mass activity, full consideration should be given to environmental capacity in the development of new tourist attractions. For instance, in rebuilding the Pavilion of Prince Teng, China extensively enlarged the land around as required; to receive 700,000 international tourists and 1,000,000 domestic pilgrims during Islamic pilgrimage period from November to December every year, Saudi Arabia has built King Abdulaziz International Airport, the waiting hall of which covers 1.5 million m^2, and the main hall is as large as 80 football fields. The airport itself is a peculiar tourism scenery.

5.5 Earthscientific Characteristics of China's Tourism Resources

As a large country in the world, China has complicated geographical environments, many ethnic groups and time-honoured cultures. It is rich both in natural tourism resources and in cultural tourism resources differing with those of other nations and countries. All these are fundamental treasures for tourism development and good for forming China's tourism advantage. Following is a general analysis of the earthscientific characteristics of China's tourism resources.

5.5.1 Regional Distribution of Mixed Banded and Blocky Tourism Resources Which Are Sparse in the West but Dense in the East

China's tourism resources are dense and elegant in the east but sparse and rugged in the west, integrating banded and blocky distribution characteristics. Karst landscapes are concentrated in South-west China; desert landscapes are mainly located in North-west China; beautiful waters and mountains appear mostly in the middle and eastern parts south of the Yangtze River; and tall mountains scatter in the west. Han culture is centred in Central and East China, while cultures of ethnic groups are concentrated in remote areas. All these belong to blocky distribution. The seaside tourism resources in the coastal areas in East China, mountain and water tourism resources along the Yangtze River, and historical and cultural tourism resources including the Great Wall and "Silk Road" are all distributed like bands. These bands and blocks interweave and interpenetrate with each other, forming multilayered combination space. Many people have brought forward schemes for China's tourism regionalization. The author holds that the key to China's tourism regionalization lies in combinations of blocks and bands.

5.5.2 Scenic Combination of Natural and Cultural Tourism Resources

"Most famous mountains are occupied by the monks". Buddhist and Taoist temples are built in almost all famous mountains in China, which become famous because they were frequented by emperors, celebrities, famous men of letters, and eminent monks and Taoist priests. The areas where mountains join plains and rivers mix with lakes are often political, economic and cultural centres or strategic positions. Beautiful natural environment together with massive buildings constitute tourism destinations of various combination types, such as gate cities, royal palaces, temporary imperial palaces, gardens, imperial mausoleums, temples and towns. Lots of natural and cultural tourism resources gather in the junction of Mount Taihang–Mount Yanshan and North China Plain, forming the 1 km gorgeous "*Golden Tourism Belt in Front of Yanzhao Mountain*" (Guo Kang 1986), which is a typical scenic combination of natural and cultural resources. Cultural factors even penetrate into some seemingly pure natural tourist attractions. For example, in Stone Forest of Lunan, besides such artificial cliffside inscriptions as "Number One Grand Spectacle on Earth", "Stone Forest" and "Baffle of the South Heaven", there are also variously shaped rocks, which are described in mysterious but beautiful myths as a sentimental and lively tourist attraction. At present, theorists emphasize the need to impart more scientific knowledge to tourists, which is a necessity as well as the development trend of high-level tourism in the future. However, vivid description of images and interesting myths are also indispensable, because what tourists primarily pursue is the spiritual enjoyment of the wonders and beauty of nature. As the formation principles of natural phenomena of the same type are almost the same, repeated explanations are boring, but the diversified shapes of imaginative landforms, unpredictable natural phenomena and mysterious legends offer people aesthetic pleasure. Therefore, only by skilfully combining science with artistic appeal and integrating scientific knowledge into tourism will the charm of tourism be inexhaustible.

5.5.3 Long-Distance Combination and Feeling and Scenery Blended Group Combination Characteristics of Famous Mountains and Gardens

Long-distance combinations of tourism resources are common in China. Besides the "five famous mountains" and "four Buddhist holy mountains" mentioned above, long-distance combination also exists in "four ancient building complexes" (The Imperial Palace in Beijing, Chengde Imperial Summer Resort and Eight Outer Temples, Confucius Temples, Confucius Mansion and Confucius Cemetery in Qufu, Shandong, and the Old Palace in Shenyang), "three famous grottoes" (Dunhuang Mogao Grottoes, Datong Yungang Grottoes and Luoyang Longmen Grottoes), "three famous towers" south of the Yangtze River (Yueyang Pavilion, Yellow Crane Tower and Pavilion of Prince Teng), "four forests of Steles" (Forest of Steles in Xi'an, Jiaoshan, Huangting and South Fujian) and ten scenic spots, sixty-two famous historic and cultural ancient cities, etc.

Chinese people have long been fond of combining the scenic spots of an area and name them collectively with exquisite words, e.g. "two scenic masterpieces in Suzhou" (gardens and water alleys), "three wonderful sceneries in Jinan" (Daming Lake, Baotu Spring and Thousand-Buddha Mountain), "four wonders of Yandang Mountain" (peaks, waterfalls, caves and stones), "three tablet wonders of Northern Wei Dynasty (386A.D.–534A.D.) in Shandong" (Zheng Daozhao's Stone Tablet of Northern Wei Dynasty in Linglong Mountain and Stele of Zheng Wengong in Hepingdu, Yexian), "three treasures in Kunshan" of Jiangsu (Kun Stone, Rare Flower and Twin Lotus), "four scenes in Xichang" of Sichuan (pines, wind, water and moon) and "eight great sites in the Western Hills" of Beijing (Temple of Eternal Peace, Temple of Divine Light, Three-Hill Convent, Temple of Great Mercy, Temple of Dragon King, Temple of the Fragrant World, Cave of Precious Pearls and Temple of Buddhahood). These not only represent a high degree of artistic combination and generalization of tourism resources, but also play a role in attracting and inspiring tourists, so as to encourage them to travel farther.

5.5.4 Characteristic of Gathering of Numerous World Wonderful Tourism Resources

China has numerous tourism resources famous for world-unique wonderfulness, and many places of interest are recognized as unique or No.1 in the world. For example, Beijing is reputed as "the capital having flourished for 3000 years in the region of latitude 40° north", "a World Wonder" and "the Zenith of the great civilization"; the Terra-Cotta Warriors and Horses of the Qin Dynasty (221B.C.–207B.C.) is called "the world's largest underground military museum"; the Great Wall is called "one of 'the seven wonders of the world'"; and Tibetan Plateau is called "roof of the world".

Many of China's tourism resources top others in the world, e.g. the world's highest peak—Everest (8848.13 m); the world's deepest canyon—Tiger Jumps Canyon (as deep as more than 3000 m); the world's oldest open-shoulder bridge—Zhaozhou Bridge (also known as Anji Bridge); the world's largest square—Tian'anmen Square; the world's longest ancient bridge—Anping Bridge; the world's grandest palace complex—The Imperial Palace in Beijing; the world's largest royal garden—Chengde Imperial Summer Resort; the world's largest stone Buddha statue—Leshan Giant Buddha; and the world's largest iron lion—Cangzhou Iron Lion. All these are monopolistic tourism resources.

5.5.5 Characteristic of Significant but Adjustable Seasonal Variations

Most areas of China are in the temperate zone and are conditioned by monsoon climate, resulting in great seasonal variations in tourism resources and evident high season and low season for tourism. The high season for some places lasts for only two to three months, causing great difficulty in the management and development of tourism. However, the high seasons for some of China's tourism resources can be adjusted and prolonged.

1. China features four distinct seasons rather than instant change between winter and summer; therefore, we can make the best of tourism resources in spring and autumn and step up stimulating and inducing tourism awareness in shoulder reasons, so as to prolong the high season for tourism.
2. China is especially rich in cultural tourism resources, which are little affected by season, such as ancient buildings, cultural relics, culture, sports and arts. Some conferences and trade fairs may be staged in low season, so may some ethnic tourism. Effective development may greatly change the desolate situation of low season for tourism.
3. Since China is vast in territory and the climate differs radically between the north and the south, large-scale adjustment of high season and low season can be made throughout the country. The emphasis of summer half year may be put on Central China and that of winter half year on the north and south, with the former launching activities centring on water resources and the latter centring on ice and snow resources.

References

Guo Kang, J. (1986). *Discussion on developing golden tourism belt in front of Yanzhao Mountain*. Geography and Territorial Research, 2(3), 45–49.
Mingde, L. (1988). *Tourism geography* (p. 12). Xi'an: Northwestern University Press.
Yunting, L. (1988). *Modern tourism geography* (p. 101). Nanjing: Jiangsu People's Publishing House.
Zhenli, L. (1987). *Chinese tourism geography* (p. 17). Tianjin: Nankai University Press.

Chapter 6
Principles of Aesthetic Appreciation of Sceneries

6.1 Definition of Scenic Beauty and Course of Aesthetic Appreciation of Sceneries

Scenery is defined both in broad and narrow senses. Scenery in broad sense refers to all beautiful natural landscapes or environments in which natural and cultural landscapes combine with each other, e.g. views hundreds of miles away, bridges and rivers ahead of courtyards; scenery in narrow sense usually refers to traditional scenic areas consisting of "famous mountains and renowned waters" in China. In general, such scenic areas refer to territorial spatial complexes which are based on beautiful and typical natural landscapes, blended with beautiful cultural landscapes, set in good environments and aimed mainly at meeting the spiritual and cultural needs of people. Sceneries with aesthetic feeling refer to natural landscapes with high natural aesthetic value; and "typicality" refers to the typicality of natural landscapes such as typical landforms including karst, Danxia and granite, which are of scientific value. Such natural landscapes should be the foundation or main body of scenery. Cultural landscapes are beautiful because they have aesthetic, historical and cultural values. For scenic areas, cultural landscapes play the supporting role as guests. Penetration means that cultural landscapes should be in harmony with natural landscapes and the order of host and guest should not be reversed. Good environment means that a scenic area should have high-quality eco-environment. Meeting the spiritual and cultural needs of people is the main function of scenic areas, which are sites of spiritual and cultural activities with various functions such as sightseeing, scientific research and popularization, creation of landscape culture, education on patriotism and religion, instead of free markets for cutting into mountains, felling trees or conducting businesses. In a word, scenic area is a large territorial spatial complex consisting of all the above elements.

© Springer-Verlag Berlin Heidelberg and Science Press Ltd. 2015
A. Chen et al., *The Principles of Geotourism*, Springer Geography,
DOI 10.1007/978-3-662-46697-1_6

Among various functions of scenic areas, aesthetic appreciation is the most important, which is the time-honoured traditional characteristic of China's scenic resorts and the reason why China's scenic resorts are of great aesthetic value. Tracing the history of aesthetic appreciation of scenic resorts may help understand the unique aesthetic values and standards of scenic resorts in China.

6.1.1 Aesthetic Appreciation Generated from Nature Worship (Xie 1987)

In ancient times with low productivity, on the one hand, people were fond of and dependent on the natural environment in which they lived; on the other hand, they feared of powerful natural forces such as high mountains, great rivers, floods and wild animals, and owing to lack of scientific knowledge, they felt mysterious about natural phenomena such as wind, cloud, rain and water and thought that all these were caused by gods, which gave rise to nature worship. The more than 400 mountains and 300 rivers recorded in the *Book of Hills and Seas* were worshiped as fairy mountains and holy waters. Rulers made use of nature worship to strengthen their governance. According to *Historical Records*, "The emperor worships famous mountains and great rivers all over the country, three excellencies worship five famous mountains and princes worship Yangtze River, Yellow River, Huaihe River, Ji River and other great rivers and famous mountains within their jurisdictions". As a result, national and local famous mountains and great rivers were recognized, so were scenic resorts. Many famous mountains and great rivers such as "the Five Famous Mountains" have now been developed into national key scenic resorts.

During nature worship, people's experience of and emotions towards natural images are linked to some extent, so certain aesthetic appreciation is involved in worship. The signs on the pottery Zun unearthed at Lingyanghe site of Dawenkou Culture have the shape shown in Fig. 6.1. According to archaeologists, Fig. 6.1a stands for the sun, Fig. 6.1b for the moon or cloud and Fig. 6.1c for mountain, indicating the worship of ancient people towards high mountains linked with the sun and the moon and their experience of the beauty of high mountains. Aesthetic appreciation is also involved in ancient emperors offering sacrifices to the heaven and earth. For example, after ascending Mount Taishan, Emperor Wu of Han complimented, "Mount Taishan is so tall, ultimate, huge, magnificent, outstanding, awesome and confusing". Except "confusing", the other six words are all praises to the magnificent image of Mount Taishan.

In the spring and autumn period, the aesthetic appreciation of famous mountains and great rivers generally appeared in the form of "comparing with virtue", that is linking some morphological attributes of natural landscapes with people's inner virtues. Confucius said, "The wise find pleasure in water; the virtuous find pleasure in hills. The wise are active, the virtuous are tranquil", which means that the limpidity of water symbolizes wisdom, and the flow of water represents the

Fig. 6.1 The signs on
the pottery Zun unearthed
at Lingyanghe site of
Dawenkou Culture

exploratory spirit of the wise, while the stillness of mountain resembles the honesty of the virtuous. That is to say, the water nourishes all creatures; the mountain invigorates all things and does favour for others, which is exactly the quality of the virtuous. It can be seen that famous mountains and great rivers have been endowed with a new function of aesthetic appreciation, though in the form of "comparing with virtue", in addition to simple primary worship.

6.1.2 Formation of Aesthetic Appreciation of Sceneries

As an object of aesthetic appreciation, sceneries are formed with the formation of scenic resorts. During the Wei, Jin and the Southern and Northern Dynasties, the Western Jin Dynasty (265 AD–316 AD) moved its imperial court to the south owing to social disorder, and the privileged nobles and men of letters lived a peaceful and comfortable life of sightseeing in the picturesque south of the Yangtze River. Those politically frustrated men of letters and eunuchs would "go sightseeing at will" or live in seclusion in the countryside to abandon themselves to nature. An example of the former is landscape poet Xie Lingyun of the Southern Dynasties, and one of the latter is great poet Tao Yuanming of the Eastern Jin Dynasty (317 AD–420 AD). Their landscape and pastoral poems have totally shaken off the form of nature worship and achieved harmony between nature and man. The landscape poems of Xie Lingyun fully reflect his unprecedented contributions to developing the function of aesthetic appreciation of scenic areas and discovering scenic aesthetic resources. The poem of Tao Yuanming, "Within the world of men I make my home, yet din of horse and carriage there is none; you ask me how this quiet is achieved, with thoughts remote the place appears alone; while picking asters beneath the eastern fence, my gaze upon the southern mountain rests; the mountain views are good by day or night, the birds come flying homeward to their nests; a truth in this reflection lies concealed, but I forget how it may be revealed", indicates that he had a close relationship with beautiful sceneries and reached the state of feeling and setting being happily blended while living in famous mountain and keeping farming and reading. So he wrote, "Life is short, why not keep happy?". His appreciation of idyllic sceneries with mountains and rivers and his imagination of Peach Blossom Land had a profound influence on the aesthetic image of famous mountains in later generations.

Coinciding with the emergence of landscape poems, landscape painting came into being as an independent painting form. Chinese landscape painting holds an important position in the history of painting of China and the world both for its long history and for its remarkable artistic achievements. "The time of its emergence as an independent painting form is more than a thousand years earlier than that of European landscape painting (Zhang 1987)". Landscape painting is the result of painters' aesthetic appreciation activities in nature. Landscape painter Zong Bing of the Southern Dynasties "travelled to Jingshan and Wushan Mountains in the west and climbed Mount Heng in the southshan", "lived in seclusion for thirty years", and "everything that is interesting can make him think…just for the ease of the mind". "Ease of the mind" refers to the spiritual comfort gained in appreciating the beauty of nature (Yang Xing: *Aesthetic Expedition of Mount Taishan*). According to *A New Account of Tales of the World*, after Gu Changkang returned from Kuaiji, people asked him about the beauty of landscapes. Gu said, "Hundreds of rocks compete, thousands of gullies vie. All the rocks and gullies are covered with luxuriant trees and grass, resembling rising rosy clouds." It shows that famous mountains and great rivers were not objects of nature worship in their eyes any more, but objects of sightseeing, residence and aesthetic appreciation. That is to say, the main function of scenic resorts has transformed from awed worship to sightseeing, residence and aesthetic appreciation.

The contributions of religious activities to the development of scenic areas are also noteworthy. Buddhism was introduced into China in the Eastern Han Dynasty (25 AD–220 AD). In the initial phase, missionary work was primarily translation of Buddhist scriptures, done mostly in cities. In the Southern and Northern Dynasties, Buddhism spreads to the whole society. In particular, beautiful famous mountains and great rivers have become ideal places for monks devoting themselves to practicing Buddhism. Taoism, founded in the Han Dynasty, was guided by the principles of loving nature, recovering simplicity and returning to nature. Taoist priests are keen on picking herbs in mountains, making pills of immortality and cultivating the heart and nature so as to "become immortal". Therefore, famous unearthly mountains detached from secular life have naturally become their ideal residences. Religions enter scenic areas not to worship nature but to take the beautiful and excellent scenic areas as places for cultivating heart and nature and carrying out religious activities. Besides, temples and Taoist monasteries offer accommodations to men of letters, tourists and pilgrims in their tours. Though monks, Taoist priests, poets and painters have their respective aspirations, they share a common language in philosophy and aesthetics, which tie them as friends in scenic areas. Among them are many eminent monks of high cultural literacy who are proficient not only in Buddhist philosophy but also in poetry, calligraphy and painting, e.g. Dao An, Hui Yuan, Fa Xian and Dai Kui. Many poets and painters such as Xie Lingyun, Zong Bing, Yan Liben, Gu Changkang and Wang Xizhi are also proficient in Buddhist philosophy and fond of making friends with monks. They are all keen on the development and aesthetic appreciation of scenic areas. It shows that persons of high cultural literacy including skilful craftsmen have from the very beginning directly participated in the development of scenic areas and their natural aesthetic resources in China. With exquisite craftsmanship, they have created colourful landscape cultures, including landscape poems, landscape paintings, landscape buildings and religious cultures such as temples, monasteries, houses, grotto images, frescoes and sculptures, which have become precious treasures of scenic resorts.

6.1.3 Development of Aesthetic Appreciation of Sceneries

The Tang and Song Dynasties witnessed rapid development of scenic areas and aesthetic appreciation in China. In the heyday of the Tang Dynasty, the prosperous economy, stable society, unified nation, developed culture and prevalent religion provided material and cultural conditions for unprecedented development of scenic resorts. Many of the existing scenic resorts had become famous as early as in the Tang Dynasty. Since the Tang and Song Dynasties, going sightseeing has become the order of the day, especially for those scholars and men of letters who "write poems and articles to express their deep feelings of the scenes and inspire the posterities", which led to vigorous development of landscape culture, travel notes, poems, paintings, gardens and landscape bonsai relating to famous mountains and great rivers. Remarkable achievements have been made in the development of natural beauty both in depth and breadth. Li Bai, who "loves going sightseeing in mountains in his life", wandered about with the lofty sentiment of "having great capacity for drinking and poetry, I take all the mountains and rivers as my home". With unique aesthetic feeling of famous mountains and great rivers, fertile imagination and romantic sentiment, Li Bai reached an unprecedented level in aesthetic appreciation of natural landscapes. All the mountains and rivers he had been to witnessed his aesthetic appreciation, for example "Its torrent dashes down three thousand feet from high, I wonder if it's Milky Way that falls from the sky" was his compliment of the waterfall of Xianglu Peak of Mount Lushan; "This morning, I depart the town of Baidi, engulfed by vibrant clouds; I return to Jiangling which was one thousand li away just in one day! The screams of monkeys on either bank had scarcely ceased echoing in my ear; when my skiff had left behind it ten thousand ranges of hills". was his aesthetic feeling towards Three Gorges on the Yangtze River. With feeling and setting happily blended with each other, this poem fully demonstrates the "ease of the mind".

Given the aesthetic pleasure of the magnificence of famous mountains and great rivers brought by his landscape poems, Li Bai seems to be the embodiment of natural beauty of famous mountains and great rivers. By contrast, Liu Zongyuan was good at seeking natural beauty in the quiet and lovely hills and streams and cultivating his mind in tranquillity. He said, "My eyes converse with the cool and refreshing air, my ears with the murmuring water, my spirit with the indifferent and detached environment, and my heart with the serene and sedate atmosphere", which theoretically explains that natural landscapes not only have visual and auditory effects, but also have spiritual and psychological functions. In Liu Zongyuan's eyes, the natural beauty of mountains and rivers was the result of the law of their natural development. His travel notes have great influence on the development of aesthetic appreciation of natural landscapes in later generations.

Su Dongpo was degraded repeatedly in his life and travelled about to secure an official position, but he loved the landscapes of his nation and sang high praise to the beauty of mountains and rivers everywhere. He "loved the steep rocks inside the gorges; liked the even ground outside the gorges; felt easy with peaceful mood; and stayed comfortable wherever he was". His poem "The Sun plays on the floodings, the scene is riant; the drizzle misting the mountains is also a sight; comparing West Lake to 'West Beauty', both are charming in light dress as well in bright" was his artistic achievement at the zenith of aesthetic appreciation of the West Lake. Besides his great contributions to aesthetic appreciation of sceneries, he also led the masses to dredge and build three west lakes, namely Hangzhou West Lake, Yingzhou West Lake and Huizhou West Lake. Today's Su Causeway in Hangzhou West Lake and Sugong Causeway in Huiyang West Lake were relics of those days. He not only went sightseeing in the famous mountains and great rivers, but also contributed to the development of landscape science which focuses on the research of beauty, as is exemplified by his *Travel Notes to Shizhong Mount*.

In appreciating natural landscapes, landscape painters have made more profound summaries of the observational, analytical and aesthetic theories relating to the features of natural landscapes. Landscape painting master and theorist Guo Xi of the Song Dynasty (960 AD–1279 AD) was the most representative. With subtle observation unique to painters, he generalized a series of methods of appreciating mountains. As he wrote in *Landscape Instruction*, "The mountain looks like this at short range, but it looks different several miles away and it changes again tens of miles away, so the shape of the mountain changes along with your paces…the mountain looks like this in spring and summer, but it looks different in autumn and winter, so the sceneries of the mountain are different in different seasons. The mountain looks like this in the morning, but it looks different in the evening and it changes again in cloudy days, so the scenes of the mountain are different at different times. As such, one mountain has the sceneries of dozens of or nearly a hundred mountains, which really deserves our deep research (Yu 1960)". That is the way of appreciating landscapes summarized by him after travelling to numerous mountains and rivers, namely the comparative and dynamic aesthetic appreciation. Based on this, he came to an important conclusion that "you know the shape of mountains and rivers from a distance and get their essences at short range". The "shape" refers to the macroscopic shape of mountains, namely the different shapes of various of mountains; and the "essence", in modern science, refers to the properties of mountain rocks and related weathering features, namely the essential features of natural landscapes. Guo Xi also explained the corresponding relationship between man and nature, saying "In spring, the mountain is cloudy, making people feel relaxed; in summer, the mountain is luxuriant, making people feel comfortable; in autumn, the mountain is clean, making people feel solemn; and in winter, the mountain is dark, making people feel lonely".

Shi Tao, a landscape painter of the early Qing Dynasty, travelled to numerous mountains in his life and lived in Mount Huangshan off and on for 23 years. Insisting on "getting inspirations from various strange peaks and drawing sketches as many as possible", he not only created plenty of distinctive landscape paintings, but also wrote many wonderful essays concerning aesthetic appreciation of landscapes.

The prevalence of religion and sightseeing since the Tang and Song Dynasties promoted the development of scenic resorts. Besides temples and Taoist monasteries, academies, libraries, cottages, places for seclusion, etc., also appeared in scenic resorts, forming important components of cultural landscapes. For example, Bai Juyi built a thatched cottage at the foot of Xianglu Peak of Mount Lushan to "view the mountains overhead, listen to the springs on the ground and watch the bamboos, trees, clouds and rocks aside from morning to evening", enjoying the visual and auditory beauty of natural landscapes. The cultural boom of the Tang and Song Dynasties gave rise to large numbers of academies in scenic areas, e.g. Yehou Academy in Hengshan of the Tang Dynasty and "Four Great Academies" of the Song Dynasty (960 AD–1279 AD) comprising Yuelu Academy, Songyang Academy, Bailudong Academy and Suiyang Academy, which inject cultural blood into scenic resorts and add new functions to them.

With the extensive development of sightseeing and deepening of aesthetic appreciation, aesthetic appreciation was integrated with landscape science. Outstanding scientist Shen Kuo of the Song Dynasty, also a traveller having visited numerous places, "studied every place he arrived". In visiting Yandang Mountain, in the face of its unique natural landscapes "Different from other mountains, in Yandang Mountain, most of the peaks are steep, dangerous and strange, towering nearly one thousand feet; and there are also high cliffs and huge valleys. All of them are circled by surrounding valleys. You see nothing if you stand outside the mountain, but you can see all of them inside the mountain", he analysed, "Owing to the impact of floods at that time, the soils and sands were all rushed away, only huge stones were left. Scenes like Dalongqiu and Xiaolongqiu Waterfalls, water curtain and New Moon Valley are all the results of water erosion". In geographer Zhu Kezhen's opinion, Shen Kuo's analysis of water erosion on landforms was 600 years earlier than the creation of erosion theory in European academic circles.

Xu Xiake made unprecedented contributions to scientific research and evaluation of scenic resources. He not only studied the characteristics of natural beauty of landscapes, but also tried to find a scientific explanation. The *Travel Notes of Xu*

Xiake, to which he devoted his whole life, is a magnum opus of landscape aesthetics and literature as well as the earliest document concerning karst landforms in the world.

Wei Yuan, a scholar of the Qing Dynasty who "set foot in almost all areas", came up with the idea of "learning from travelling amid mountains". In *The Song of Travelling amid Mountains*, he wrote "People just know the fun of travelling amid mountains, but not know the meaning of it; born in nature, can the breath of man not connect with nature? Trees form the boundaries of the mountain, caves make the mountain clever, still water makes the mountain mysterious, and running water connects all mountains together; springs make mountains quiet and rocks make them magnificent; clouds make the mountains alive and trees make them luxuriant; who can realize the truth and become wise without exploring the springs, stones, clouds and trees in nature? One sees the surface of the mountain if he takes an extensive tour; but one sees the soul of the mountain if he takes an intensive tour; only when you become united with the mountain can you really get to know it. Animals all envy our human beings; let them enjoy the clouds, mountains and sunset with us". Wei Yuan made a dialectical analysis of the function changes of natural landscapes such as mountains and the corresponding relationship between men and the shapes of sceneries. "Secluded mulberries may hide themselves in remote places and deep mountains are also wonderful"; "peculiar sceneries are always in precipitous places and pleasure is always obtained from arduous exploration"; "I favour the peculiar, precipitous and secluded landscapes, the pleasure of which is known by no one except me; scenery which is not deep, secluded, profound or vast is the dullest". That is the aesthetic appreciation achieved by means of comparison. Only in this way can one feel "the benefits of one tour are more valuable than ten years of reading".

6.1.4 Deepening of Aesthetic Appreciation of Sceneries

With the end of the long feudal regime and the development of modern science and culture, scenic resorts in China are developing towards more functions such as tourism, aesthetic appreciation, scientific research, popularization of science, vacation, education and religion. In particular, with the development of modern natural sciences, scenic areas, as beautiful spaces integrating natural relics in the earth's history and cultural relics in human history, have naturally become sites of scientific research and education popularization, e.g. landscape geology, landform, hydrology, vegetation, animals, climate, and humanities including archaeology, history, aesthetics, culture, art, architecture and tourism. Therefore, aesthetic appreciation of natural landscapes is increasingly more closely linked with natural sciences, and aesthetic appreciation has become colourful, multifold and stereoscopic.

Since the founding of the People's Republic of China, especially in recent years, scenic resorts have been restoring, developing and improving at a speed never seen before in history. Diversified, colourful and distinctive systems of scenic resorts in China are like precious jades embedded in the beautiful land, and with their special charms, will be enjoyed by future generations.

6.2 Characteristics and Structures of Natural Beauty of Sceneries

6.2.1 Characteristics of Natural Beauty of Sceneries

1. Nature, namely objectivity. Objectivity is the fundamental characteristic of the natural beauty of sceneries, and it is endowed by nature and independent of man's will. People can feel and appreciate its extraordinary beauty, but they can never create it, e.g. the magnificent Mount Taishan, precipitous Mount Huashan and beautiful Mount Emei. Therefore, the objectivity of natural beauty of sceneries is the material base of beauty of sceneries as well as one of the fundamental characteristics making it different from social aesthetics and artistic aesthetics.
2. Infinity of scenic space. Different from other objects of aesthetic appreciation, sceneries are unbounded nature. Infinity refers to the wonderful feeling of one's mind running free in the infinite nature. For example, at the top of Mount Taishan, there is Sunrise Peak to watch the sunrise on the East China Sea, Yue Peak to watch the landscapes of Wu and Yue in the south and Qin Peak to watch the 800-li Qinchuan in the west. Of course, this was the intuition and imagination of the ancient people, but it also indicates that one can see as far as the eyes can reach and set his mind free in the infinite space when climbing mountains.
3. Hierarchy of scenic space. Infinite scenic space is made up of countless levels of sceneries. Any scenic spots in scenic areas have countless levels, from near to far, low to high or shallow to deep. That is the so-called theory of level distance, deep distance and high distance. For example, in Mount Huangshan, when you look up from Peach Source Pavilion to Yuping Peak and Tiandu Peak, you can see trees, stones, mountains, mountain ranges, peaks, clouds, etc., in the infinite universe, which are sceneries of high distance; in Three Gorges on the Yangtze River, the rushing water crossing mountains and valleys and the mountain ranges separated by the Yangtze River are sceneries of deep distance; and "his lessening sail is lost in the boundless blue sky, where I see but the endless River rolling by" is scenery of level distance. Though sceneries belong to different areas, such areas are not boundaries of the objects of aesthetic appreciation.
4. Synthesis of scenic space. Infinite scenic space is a territorial complex consisting of various organic and complex natural scenes. It is an epitome of the organisms in the space of the earth surface, including lithosphere, hydrosphere, biosphere, atmosphere and even the sun, the moon and the stars in the universe.

They cross to form special scenic landscapes in a specific geographic position. In China, these special crossed areas form many colourful complexes, which have become distinctive scenic areas after thousands of years of selection, development and protection by the Chinese nation.

5. Sceneries are spaces of omnidirectional aesthetic appreciation. As objects of aesthetic appreciation, sceneries are different from other objects of aesthetic appreciation because the subjects of aesthetic appreciation of sceneries are humans who are always integrated with the sceneries. Humans and objects not only have the relationship of aesthetic appreciation, but also are organically connected in ecology. For other kinds of aesthetic appreciation such as aesthetic appreciation of landscape paintings, films and dramas, the subjects are outside the objects of aesthetic appreciation.

Integrated with the sceneries, man can look up to the sky or look around, can sit to see the sunrise and sunset clouds or walk to see mountains and rivers. Owing to the infinity and synthesis of scenic spaces, the best way of aesthetic appreciation of sceneries is to walk around to enjoy the beauty of sceneries. Since scenic areas have aesthetic, scientific, historical and cultural values, subjects of aesthetic appreciation can gain various benefits in moral qualities, intellectual ability, physical fitness and aesthetic appreciation in the process of aesthetic appreciation while wandering about and blending your feelings with the settings.

6.2.2 Structure of Natural Beauty of Sceneries

Aesthetic feeling and appreciation are both subjective consciousness which varies from person to person. Aesthetic feelings and appreciations are always different among people of different ages, classes, nations and with different levels of education and even experiences. However, from the macro-perspective, people have some aesthetic feelings which are basically the same. For instance, almost all people, ancient or modern, Chinese or foreign, who have been to Guilin agree that "Guilin has the best scenery in the world". It shows that people have something in common regarding the understanding of the natural beauty of sceneries. So, the Chinese nation with 5000 years of civilization have much more in common regarding the understanding of the natural beauty of sceneries and have formed some traditional views on the natural beauty of sceneries.

Natural beauty of sceneries includes image beauty, colour beauty, line beauty, dynamic beauty, static beauty, olfactory beauty and auditory beauty, among which image beauty is the core and foundation of the beauty of natural sceneries.

6.2.2.1 The Image Beauty of Natural Landscapes

Image beauty refers to the beauty of overall shape and spatial form of natural landscapes. It includes people's psychological and physical feelings towards various natural landscapes. According to the evaluations made by ancient and modern poets, painters, travellers and tourists on the natural beauty of China's traditional famous mountains and great rivers and on the basis of modern earthscientific investigations, the image beauty of natural landscapes can be summarized as magnificence, peculiarity, peril, elegance, seclusion, profoundness, vastness and wildness. These characteristics are formed by scenic elements in different geological and geographic environments. Therefore, in appreciating the beauty of natural landscapes, one should start from an analysis of image characteristics, grip the essential features of scenic elements and study them according to different geological and geographic conditions.

Magnificence What landscape is "magnificent"? The magnificence of a mountain mainly refers to its lofty image. Height includes relative height and absolute height, with the former relating to the magnificence of landscapes. Standing on the eastern edge of the extensive North China Plain, Mount Taishan towers over the Qilu Hills with its great relative height, so it is famous for "magnificence". Mount Emei is known as "the most magnificent and elegant in the South-west". The "magnificence" also refers to its height, as is described in "rising by 3000 m from the ground". Besides, steep mountain slope and erect outline can also present a magnificent image. For example, a mountain with a height of 100 m and a gradient of 60° looks more magnificent than a mountain with a height of 1000 m but a gradient of 20°. That is because the greater the gradient is and the bigger the elevation angle of eyes is, the more magnificent the mountain will be.

Peculiarity What landscapes are "peculiar"? Peculiarity means unique and unexpected shape of mountains. Mount Huangshan is the most wonderful under heaven with its unique mountains, rocks, pines, clouds and springs. For example, peculiar peaks of Mount Huangshan roll one after another and tower into the sky, and there are 72 peaks higher than 1000 m, scattered randomly and presenting unpredictable landscapes; peculiar rocks of Mount Huangshan are unique and have endless shapes; peculiar pines of Mount Huangshan, which are luxuriant and straight, cling to precipitous rocks or come out of the rocks among cliffs; the sea of clouds of Mount Huangshan is enveloped in mist and clouds and rises and falls like waves of the broad sea; and the springs of Mount Huangshan keep gushing and never dry up or overflow, providing perfect conditions for shower and refreshment for tourists. Other examples are Guilin scenery with peculiar peaks and caves, Danxia landform (in Mount Wuyi), which is the "most peculiar and elegant in South-east China", which are all peculiar compared with normal landforms.

Peril What landscapes are "perilous"? From the perspective of the shapes of mountains, huge gradient, tall and narrow ridges usually form perilous mountains. As a popular saying goes, "There is only one road leading to Mount Huashan", Mount Huashan is so perilous precipitous and perilous that it looks just like a huge pillar among the peaks in front of Qinling Mountains if you have a bird's eye view of it. The cliffs of Mount Huashan are extremely steep with a gradient of 80°–90°. The fall between the top and the bottom of the peak is over 1000 m, and its main peak is about 2100 m high. To get to the top of Mount Huashan, tourists have to climb by holding the iron chains and pass through dangerous paths such as "Thousand-Foot Precipice", "Hundred-Foot Crevice", "Ear Touching Cliff" and "Ladder to Heaven" before enjoying the infinite scenes at the top. Besides bringing aesthetic experience to tourists, perilous mountains can also inspire people's curiosity of adventure and the resolution to conquer these perilous mountains. Climbing perilous mountains is also a process of annealing willpower, so perilous peaks are always the biggest attractions of famous mountains. It can be said that "every famous mountain has its perilous peaks".

Elegance What landscapes are "elegant"? Elegant sceneries must meet the following requirements: firstly, the mountains are covered with luxuriant vegetation without few bare rocks and lands, presenting a green and lively picture; secondly, the mountain's shape is unique and plump with tender and beautiful outlines. There will be no luxuriant flowers, plants and trees but for water, so elegant mountain sceneries often go with waterscapes. The phrase "beautiful mountains and clear waters" explains the scenic relationship between waters and mountains. Blessed with plenty of rain and thick vegetation, many famous sceneries in South China are known for their elegance, e.g. the magnificent elegance of Mount Emei, delicate elegance of West Lake, splendid elegance of Fuchun River and the peculiar elegance of Guilin and Yandang Mountain.

Seclusion What landscapes are "secluded"? Secluded sceneries are often represented by semi-enclosed spaces based on landforms of towering mountains, deep valleys or piedmonts and full of luxuriant and tall trees. In such landscapes, the field of view is narrow, light is not enough and the air is clear. The landscape is deep, hierarchical and full of twists and turns and so rejects a commanding view. "Seclusion" is closely related to both "depth" and "tranquillity". The famous line "A winding path leads to secluded place" explains that seclusion also contains the elements of depth and tranquillity. Such landscapes are always seen in small intermontane basins surrounded by hills on three or four sides, leaving one side as exit. These basins are often ideal places for building temples, e.g. Fuhu Temple in Mount Emei, Taihuai Town in Mount Wutai, Jiuhua Street in Jiuhua Mountain, Puzhao Temple in Mount Taishan, Small Peach Garden in Mount Wuyi and Bailudong Academy in Mount Lushan, all of which give people a feeling of seclusion.

Profoundness What landscapes are "profound"? Profound landscapes refer to those more enclosed than secluded landscapes. For example, a place ringed with mountains, with winding paths leading out along rock fissures and as profound as a well; shallow caves in precipices; and karst caves in limestone regions. Lingyan Temple, Great Longqiu Waterfall and Guanyin Cave of Yandang Mountain, Cha Cave and Water Curtain Cave of Mount Wuyi and various karst caves of limestone regions all appear to be mysterious.

Vastness What landscapes are "vast"? Vast landscapes are mainly distributed in areas with wide views, such as broad water surfaces and vast plains. Other examples are "vast expanse of waters of 800-li Dongting Lake" seen from the top of Yueyang Tower, the sight of "his lessening sail is lost in the boundless blue sky" seen from the top of Yellow Crane Tower, the panoramic view of "500-li Dianchi Lake" seen from the top of Grand View Pavilion, and "holding all mountains in a single glance" from the top of Mount Taishan. Vast landscapes make one completely relaxed and happy and thick with memories.

Wildness What landscapes are "wild"? Wild landscapes refer to natural landscapes that have not been damaged or seldom disturbed by humans, namely virgin mountains, waters, forests and lands, known as "rustic charms" in our country. Marx also stressed "wild" landscapes when he talked about natural beauty.

Ancient people made a lot of comments on the images of famous mountains. The above-mentioned honourable titles given to Mount Taishan, Mount Huangshan, Mount Huashan, Mount Emei and Qingcheng were all descriptions of respective macroscopic characteristics. Actually, every famous mountain is a colourful spatial complex of beauty comprised of the whole constituted by various scenic images which are magnificent, peculiar, perilous, elegant, secluded, profound and vast according to the rhythm and cadence of nature (such as geological evolution, landform evolution and geographic landscape change). These basic characteristics of image are not isolated but changeable and related to each other. They can compose endless movements like musical notes.

6.2.2.2 The Colour Beauty of Natural Landscapes

Nature has created not only towering mountains with a height measured by kilometres, but also exquisite stones with a length measures by inches. Besides various kinds of image beauty, nature also offers a profusion of colour beauty. The most noticeable colours are shown by colourful flowers. Many famous mountains have their unique flowers such as camellias in Cangshan Mountain of Yunnan, azaleas in Mount Emei and apricot blossoms in Badaling. The succession of the seasons and cloudy, sunny, rainy and gloomy weather conditions cause macroscopic changes of the colours of nature. The saying "emerald in spring; green in summer; gold in autumn; and silver in winter" explains the seasonal changes of natural landscapes. For example, the florid picture of ring upon ring of woods tinted with deep red as described in the saying "red leaves are as beautiful as spring flowers" in late autumn on Fragrant Hills of Beijing is not to be seen in every scenic area in every season. In nature, the colours of rock and earth are relatively stable, whereas the most colourful and changeable colours come from plants. The most common colour of natural mountain landscapes is green, which is also the most pleasant colour motif for vision. Among the many natural objects, perhaps the one with the most changeable colours is grotesque and gaudy clouds. At certain locations of some famous mountains, it is an unusual pleasure to enjoy scenes such as the morning sun rising in the east, the sky blossoming with lights, the sun sinking in the west and a great suffusion of sunset clouds. The sunrise seen from Sunrise

Peak of Mount Taishan and sunset glow viewed from Cloud Dispelling Pavilion of Mount Huangshan are both marvellous sights known far and wide since ancient times, and the miraculous "Buddha Light" at Golden Peak of Mount Emei is a wonderful view which is even more attractive.

Water is a big mirror of nature which reflects the sceneries around the sky and makes the colours of scenic areas richer and more vivid. The colours of some waters are transparent because of certain kinds of minerals. That is why the water in places like Five Colour Pool in front of Nine-village Valley and Huanglong Temple is so bright and clear. The landscape of sea embracing mountains enables people to have broad mind and surging emotions. Moreover, the blue, boundless sea and white sprays give people a profound and surging feeling. Standing on the Tide-watching Pavilion of Putuo Mountain, you will enjoy the charm of sound and colour as well as the combination of dynamics and statics.

In natural sceneries, lily-white snow is always an important material for poets and painters as well as a great attraction for travellers. The first time Xu Xiake climbed Mount Huangshan, he was greeted by the beautiful scene "The stone steps were buried by heavy snow and the path was just like white jade in a glance". The vast glaciers and snow mountains in Qinghai–Tibet Plateau in China are precious tourism resources to be developed, and therefore, they are very attractive to travellers who dare to explore the secrets of nature. In fact, some snow mountains have become borrowed sceneries of some scenic areas long ago, e.g. the snowy Minya Konka and Great Snow Mountain seen from the Golden Peak of Mount Emei are regarded as wonderful views; the enchanting Jade Dragon Snow Mountain reflected in the famous scenic area Black Dragon Pool of Lijiang, Yunnan, has been known far and wide.

Hazy clouds and thin mist always put on a harmony colour for the mountains, making the colours of mountains soft, plain and harmonious. Morning fog, just like chiffon, can conceal miscellaneous stones and branches and increase the sense of whole of mountain ranges, making ring upon ring of mountain ranges richer in levels. For example, the limestone peak forest in Guilin is more beautiful when watched from a distance, more fantastic in back light than in front light and in its best when encircled in thin mist or hazy clouds. Many famous mountains in China, especially in the south, have another kind of beauty and charm in rain, mist, morning sunshine or afterglow.

6.2.2.3 The Dynamic Beauty of Natural Landscapes

Motion and stillness are relative, opposite and complementary to each other. Both the state of people in motion with objects still and the state of objects in motion with people still will cause dynamic changes. In the past, people had the feeling, "Seen from the boat, the hills on both sides were just like running horses; hundreds of ranges of hills had slipped past just in a blink of the eyes". That kind of feeling is even stronger nowadays when one can see ranges of hills on a car, a train or even a plane. In the mountains, the waterfall is still like a sheet of silk hanging in the sky when watched from a distance; however, it is vigorous like flying dragon and dancing phoenix when watched at short range. The dynamic beauty of natural landscapes mainly consists of elements such as running water, waterfall, cloud rack and floating smoke. Running water and waterfall are relatively stable elements of natural landscapes (they also change with season). Without the huge and rolling currents of the Yangtze River, the landscape of the Three Gorges on Yangtze River will not give people a kind of exciting and soul-stirring beauty. Huangguoshu Waterfall of Guizhou, Dadieshui Waterfall of Yunnan, Dalongqiu Waterfall of Yandang Mountain and Sandie Spring of Mount Lushan all bring vitality to surrounding mountains and so become main scenic areas or spots of respective places.

Floating clouds and smoke rise slowly from deep valleys; ranges of hills dodge in and out of clouds and mist. The clouds move with wind, and the mountains seem to move with the clouds, forming a scene that "the mountains are ethereal and dreamy", which is also a kind of dynamic beauty. Clouds and smoke are unstable elements, but they also follow some laws. The general law governing their changes can be understood by a scientific analysis of the characteristics of the microclimates of scenic areas. The changes of clouds and smoke are much richer than those of running water. Sometimes the clouds and smoke are turbulent like the waves of the sea; sometimes, they are buoyant and elegant, slowly passing your feet. Being in such a landscape, you may feel that you are wandering about in a "fairyland". Though invisible, wind is also a force to produce dynamic beauty. It can drive floating clouds, raise waves, sweep willow branches, produce billowing pines, etc. In a word, dynamic beauty makes mountains and waters alive.

6.2.2.4 The Auditory Beauty of Natural Landscapes

Besides visual beauties such as image beauty, colour beauty and dynamic beauty obtained from nature, there is also auditory beauty produced by famous mountains and great rivers. People obtain enjoyment of music in special environments such as the waterfall dashing into the deep pool, the waves beating against the shore, the stream babbling in the valley, the spring pouring into the pond, the wind blowing the pine, the birds singing in the peaceful forest and the insects chirping in silent night. The Billowing Pines Pavilion, the Pavilion for Listening to Spring, etc., in some famous mountains are built for tourists to appreciate the "music" of nature. There is a special species of "music-playing frogs" in the frog pond beside Wannian Temple of Mount Emei. Every night, the frogs croak together in different pitches with a sound like that produced by stringed instruments, "playing music" for tourists who stay in Wannian Temple for the night and offering a special pleasure. Different mountains have different "sources of sound". Attention should be paid to this resource in the development of famous mountains. For those who have long lived in noisy and bustling cities, it is undoubtedly a great enjoyment to listen to the natural symphony in famous mountains and great rivers.

6.3 Cultural Landscapes Blended into Natural Landscapes

Glorious relics of ancient civilizations can be found in all famous mountains and great rivers in China. Today, people visiting famous mountains can enjoy the beauty of both natural landscapes and the cultural landscapes recording the long history of China.

As cultural landscapes are only decorative elements of natural sceneries, natural landscapes are the host, while cultural landscapes are the guest and the order of them should not be reversed. This is the most important principle in the construction of scenic areas in China.

Such cultural landscapes generally include cultural landscapes compatible with natural landscapes and the myths, legends and local customs adding extraordinary splendour to famous mountains.

6.3.1 Cultural Landscapes Compatible with Natural Landscapes

The beauty of cultural landscapes in scenic areas mainly lies in its harmony with and enhancement of the beauty of natural landscapes. Generally speaking, various buildings including sightseeing roads are the most prominent contents of cultural landscapes in scenic areas, and the less important contents are cultural relics such as cliffside inscriptions, stone tablets, calligraphies and paintings. Since artificial buildings have strong impact on natural landscapes, both the layout and design of buildings were prudently and carefully worked out in the thousands of years of construction history of scenic areas in China. Many of the existing old buildings which have been destroyed and rebuilt for several times deserve to be regarded as treasures to be kept forever. In visiting the famous mountains and great rivers, you may find that the buildings can generally blend into and enhance the beauty of natural landscapes, which is the key to architectural landscapes.

In "magnificent" scenic areas or spots, the buildings are always laid out on ridges, mountains tops or sunny slopes (bank slopes) to blend into or intensify the magnificent and towering image of the mountains. For example, on Mount Taishan where the emperors of various dynasties frequently held worship ceremonies, the buildings are well arranged and spectacular. Dai Temple was built on the alluvial fan at the southern foot of Mount Taishan, with its back leaning against Mount Taishan, front facing the plains and central axis pointing to the top of Mount Taishan; in the high and vast places along mountain roads, there are First Gate to Heaven, Halfway Gate to Heaven and South Gate to Heaven and the Jade Emperor Temple on the top of the mountain. As for the broad mountain roads, the section leading to South Gate to Heaven is like a ladder towering into the clouds or the path leading to "heaven". The buildings are surrounded by pine trees, cypress trees and Chinese scholar trees, making Mount Taishan even more majestic. There are many similarities between the layouts of buildings in Mount Hengshan (in the south) and Mount Songshan. Tiantai Peak of Jiuhua Mountain and Golden Peak of Mount Emei both occupy a commanding position, which intensifies the majestic and towering beauty of the mountains.

In "perilous" scenic areas or spots, the buildings are mostly laid out along steep cliffs. South Gate to Heaven, Chess Pavilion, "Sparrow Hawk Turning over the Body" and Sky Plank Road of Mount Huashan are all examples of this. The Hanging Temple of Mount Hengshan (in the north) was built in the recesses of rocks in step cliffs, with its beams penetrating into the cracks of rocks and pavilions connected to plank roads. The whole temple has 40 palaces which are all on the verge of cliffs or face deep pools.

In "elegant" scenic areas or spots, the buildings are mostly surrounded by hills on one side and waters on the other and are set amid forests. Such architectural landscapes are common south of the Yangtze River where the hills are green and waters are clear. Generally, pavilions delicate in style, proper in size, appropriate in dimension and simple in colour are set at the main viewing points to enable the tourists to appreciate the landscapes, increasing the elegance of sceneries. Pagodas may be built on the ridgelines of small hills on hilly lands with lakes and mountains, so as to create sceneries by breaking through the gentle curves, as is exemplified by scenic areas such as Hangzhou, Guilin and Taihu Lake.

In "secluded" scenic areas or spots, the buildings are often laid out amid ancient dense forests at mountain foots, valleys and small intermontane basins, creating a beautiful environment in which ancient temples and palaces hide in deep valleys and dense forests. Lingyan Temple (Temple of the Soul's Retreat) of Yandang Mountain, Hongchun Terrace of Mount Emei, Yongquan Temple in Gushan of Fuzhou and Tanzhe Temple of Beijing were all built in such environments.

In "mysterious" scenic areas or spots, the buildings are often integrated with caves. For example, Guanyin Cave at the foot of Hezhang Peak of Yandang Mountain is a nine-storey palace built along mountain terrain. From a distance, it looks like a natural cave, and from nearby, you may see some palaces and pavilions. Inside the cave is the grand nine-storey palace which can hold thousands of people. Buildings of this kind can also be found in Beidou Cave, Jiangjun Cave and Shiliang Cave as well as Fairy Cave of Mount Lushan.

In "peculiar" scenic areas or spots, the design and layout of buildings are usually surprising and tactful. Taking advantage of the spectacular stone forest, the sightseeing roads of Stone Forest in Yunnan rise, fall, pass caves and cross bridges randomly, winding and bewildering like a maze; following the trends of cliffs, a nine-layer pavilion was built in Shibao Stronghold of Sichuan as the stairs for climbing Yuyin Mountain. What a unique design! In Cangyan Mountain of Shijiazhuang, a bridge was built over deep ditches and steep cliffs and a palace was built on the bridge, hence the name "Bridge-Tower Hall", which is unlikely to be conceived and built by ordinary people.

In "vast" scenic areas or spots, the buildings are often laid out on hills or plateaus facing a river or a lake or in the building complexes amid forests. Main buildings are also built to make it possible for tourists to look far into the distance and have a broader field of vision. For example, Yellow Crane Tower of Wuchang, Yueyang Tower of Dongting Lake, Six Harmonies Pagoda of Qiantang River and Yugu Pavilion of Ganzhou.

Focusing on hiding but not neglecting exposure is a great feature of artistically handling the relationship between buildings and surroundings in China's scenic areas. In natural landscapes characterized by "magnificence, peril and vastness", the buildings are usually "exposed"; in natural landscapes characterized by "elegance, seclusion and profoundness", the buildings are mostly hidden, and it is common that "exposure" introduces "hiding", e.g. expose mountain gates to introduce grand palaces hidden and expose decorated archways and stone carvings of humans and animals to introduce imperial tombs (such as Ming Tombs) hidden in the deep and dense woods at the foots of mountains.

The beauty of buildings themselves mainly lies in the spatial combination of single building and building complex. The buildings on famous mountains in China are characterized by their unique ethnic forms and local styles as well as harmony with nature. Harmony with nature means integrating the layout of buildings (including vertical layout), individual design and artistic treatment of details of buildings into the environment and forming an organic scenic complex through combining the terrain of each scenic spot in scenic areas and the features of natural landscapes and making full use of natural space.

The sightseeing routes of the traditional places of interest in China were also chosen carefully. For example, built along the trends of the mountain, the incense path of Putuo Mountain has become the link connecting three temples, 36 nunneries and 128 cottages. The path was paved with flagstones and lined with various carved lotuses every ten steps. However, some sections have been turned into dusty highroads, which is really an unpleasantness.

Apart from buildings, cultural landscapes also include tablets, cliffside inscriptions, calligraphies and paintings, historical relics, revolutionary cultural relics and so on, which are all important cultural relics of China. To eulogize the beauty of nature with the beauty of literature and art can bring out the theme, polish up the landscapes and light up the scenes of scenic areas. In such a way, the tourists cannot only appreciate the beauty of sceneries, culture and art through sightseeing, but also gain historical and cultural knowledge. To integrate history, culture, art and nature into famous mountains is a highlight of scenic areas in China. For example, the words "having a panoramic view of the flying clouds" were inscribed in a giant towering stone in front of Fairy Cave of Mount Lushan. Whenever floating clouds surge around the valley and form a sea of rolling clouds, only the giant stones tower out of the clouds, presenting ineffable profoundness of enjoying "a panoramic view of the flying clouds". Such inscriptions like "the stream is clear; the mountain is clean" on the Heshui Rock (Yelling at Water Rock) in Gushan, Fuzhou and "Ultima Thule" and "Pillar of the South Heaven" of Sanya, Hainan, all play the role of "putting the finishing touch to the picture of a dragon" and inviting the ravished eyes. Poems, paintings and inscriptions of famous mountains have their own historical and cultural values, e.g. from stone inscriptions of Lisi of the Qin Dynasty to the end of the Qing Dynasty, thousands of stone inscriptions in regular, cursive, official and seal scripts have been left in Mount Taishan, turning Mount Taishan into an exhibition of calligraphies, poems and paintings.

There are also many precious cultural relics in scenic resorts, e.g. Canoe-shaped coffins (cliff tombs) dating back to more than three thousand years ago on the cliff of Mount Wuyi, which has become a wonder both today and in the past.

6.3.2   Myths, Legends and Local Customs Adding Extraordinary Splendour to Sceneries

The forms and characteristics of the beauty of natural landscapes and cultural landscapes are both related to the image and often closely linked to natural geography. For example, the landforms and hydrological phenomena such as karst caves and waterfalls are always legendarily related to Dragon King of the Sea; giant rocks fractured along joints usually produce such legends like the "Sword Testing Stone" and "Uncanny Workmanship"; and the landforms shaped like humans or animals are honoured as gods or celestial beings. It is noteworthy that landscapes with similar landforms also have similarities in their related myths and legends, which can be regarded as a reflection of traditional ethnic cultures.

Sometimes, myths and legends also have historical and cultural geographical features. Legends about Mount Taishan are always related to emperors holding worship ceremonies on mountains; legends about Dragon and Tiger Mountain in Jiangxi are often related to Master Zhang of Shangqing Palace, for this place is said to be his hometown; and the myth about "Husband and Wife Peak" in Yandang Mountain organically links the 20 lifelike strange peaks and rocks resembling humans and animals.

Beautiful myths and legends about famous mountains and great rivers subtly and harmoniously link natural landscapes with cultural landscapes like a colourful chain, making landscapes more attractive for tourists.

Different cultures and customs of various ethnic groups and the differences in local conditions of various regions are also essential to beautiful landscapes. The customs of Bai people in Cangshan Mountain and Erhai Lake, the legends of Yi people in Stone Forest of Lunan and the tropical jungles and customs of Dai people in Xishuangbanna all add charm to the famous mountains in these areas. Even in areas inhabited mainly by Han people, beautiful natural landscapes are tied up with ethnic customs. For example, "Flying across the Sky" at Southern Gate to Heaven in Lingyan of Yandang Mountain and fairs held in Zhongyue Temple of Mount Songshan may deeply impress tourists.

Aesthetic appreciation in scenic areas is achieved through sightseeing, and the best way of sightseeing is walking. Only by getting into the various aspects of beauty by climbing the mountains and crossing the rivers can one fully enjoy the beauty of sceneries. Aesthetic appreciation can achieve an integrated effect of aesthetic, intellectual, physical and moral education: aesthetic education: cultivate the mind and spirit and enjoy the beauty through appreciating natural and cultural landscapes; intellectual education: explore knowledge about natural sciences and humanities in aesthetic appreciation; physical education: it is the best fitness exercise to climb mountains and wade (including swimming by the seashore) in the beautiful and quality eco-environment; and moral education: the charming landscapes of the motherland can inspire people's patriotic sentiments and culture harmonious friendships in aesthetic appreciation.

"This land is so beautiful that it has made countless heroes bow in homage". For thousands of years, the glorious achievements made by excellent sons and daughters of the Chinese nation have added ever-lasting splendour to the magnificent landscapes of China. From the past to now, how can it not make people feel proud and excited!

This land is so beautiful that it has made countless men of letters wield their writing brushes. Famous mountains bring aesthetic mood, and beautiful sceneries inspire poetry. From ancient times to now, numerous works of landscape art have become sublime odes to sceneries, gleaming proudly in the treasury of literature and art of China.

This land is so beautiful that it has made countless scholars explore it. For thousands of years, the mysteries of various natural landscapes have attracted more and more scholars to explore them, which brings the existing scenic areas into an age of aesthetics and science and turns them into museums of the civilized society.

This land is so beautiful that it has made countless skilful craftsmen leave creations on it. Exquisite ancient buildings, immortal cliffside inscriptions, ingenious trails on precipitous cliffs, etc., all bear witness to the wisdom and artistic talents of the Chinese nation.

This land is so beautiful that it has made countless tourists go sightseeing in it. It is one of China's good traditions to worship nature and appreciate landscapes. With the development of socialist modernization, tourism and aesthetic appreciation will continue to develop vigorously and become part of the civilized life.

This land is so beautiful that all Chinese people are proud of it. Let us protect and explore more beautiful scenic areas and make the splendid landscapes in China more enchanting.

References

Xie, N. (1987). *Famous mountains in China*. Shanghai: Shanghai Education Press.

Yu, A. (1960). *Collection of painting theories*. Beijing: Beijing People's Fine Arts Publishing House.

Zhang, G. (1987). *History of Chinese fine arts*. Beijing: Beijing China Knowledge Press.

Chapter 7
Survey and Evaluation of Tourism Resources

7.1 Survey of Tourism Resources

China is vast in area and rich in tourism resources. However, for some historical reasons, not adequate attention was paid to tourism in the past, therefore considerable potential tourism resources have not been discovered and the resources already discovered have not been properly developed and fully utilized. With the rapid development of tourism in recent years, people have gradually realized that tourism resources are obviously insufficient and a systematical survey and evaluation is urgently needed to meet the needs of constant development of tourism.

7.1.1 Survey Objectives

Survey of tourism resources aims at investigating the resources that can serve tourism, so as to make preparations for the evaluation, classification and planning of the resources and lay a foundation for rational exploitation of them. It is the basic task of tourism earth-science to use earth scientific theories and methods to investigate all the tourism resources available for exploitation and provide a basis for making decisions on tourism development.

7.1.2 Survey Guidelines

General survey of tourism resources is aimed to exploit resources and ultimately serve tourism. Therefore, attracting the most tourists is the starting point of all considerations and a basis for allocating survey workload. In that light, we propose four concepts.

© Springer-Verlag Berlin Heidelberg and Science Press Ltd. 2015
A. Chen et al., *The Principles of Geotourism*, Springer Geography,
DOI 10.1007/978-3-662-46697-1_7

1. Service concept: Survey of tourism resources should centre on tourism demands and serve both existing projects and long-term strategic goals. Currently, some people who are keen on research on tourism resources still fail to break away from their professional habits formed over the years and cling to its traditional modes in their work. As a result, they have done much work dutifully, but users can neither understand nor benefit from their work. For example, some geological professionals still follow the general procedure of geological work in survey of tourism resources, and finally submit a geological report concerning strata, structures, magmatic rocks and minerals. Since the evaluation of tourism resources is rendered in abstract words, users feel the report useless, while the surveyors think that the users do not respect science. Therefore, we should change our concepts and make surveys with an awareness to serve tourism.

2. Full-range concept: Survey results are oriented to the whole tourism undertaking and should serve all tourists rather than professionals and a small number of enthusiasts. Therefore, survey of tourism resources must comprehensively reflect various factors of resources and aim at a comprehensive evaluation. Except for regions of special significance, it is unadvisable to emphasize a certain aspect alone. For example, one area of a certain seashore city has beaches and geomorphic landscapes as well as favourable traffic conditions, so it can be developed into a multifunctional tourist area. However, owing to professional habits, investigators emphasized such geological characteristics as local stratigraphic sections and ignored the potentials and benefits of full-range development. Two years' practices show that most tourists came for the beaches and landforms there, while very few specialists and scholars simply made geological and geomorphological surveys. Therefore, a wrong development direction was selected, reducing the utilization value of resources.

3. Coordination concept: Tourism is a product of social, economic and cultural development, so survey of tourism resources should be based on a consideration of coordination between local economy and society, that is coordination between a proper number of tourist spots and the distribution and supporting functions of these tourist spots. When surveying a certain tourism resource, we should also take into account the coordination between its internal functions. In that light, surveys should be made in an appropriate area to discover developable resources meeting the need of coordination between the economy and society of that area. Some people say that there is no tourism resource around cities located on plains, such as Shanghai, Harbin and Shijiazhuang, but tourism resources must be "created" to meet objective needs. Then we should survey and explore these regions to "create" tourism resources reasonably and under the most appropriate conditions.

4. Benefit concept: Attention should be paid to the following benefits in our survey of tourism resources:

   (a) Social benefit: Be able to attract and accommodate tourists, satisfy their needs and provide people with a series of bases for broadening horizons, increasing knowledge, moulding temperament and receiving spiritual civilization education;

(b) Environmental benefit: Be able to beautify and protect environments and offer people some ecologically balanced spaces and places which are beneficial for physical and mental health;

(c) Resource benefit: Be able to make full and rational use of tourism resources and give play to their potential functions and overall effects;

(d) Economic benefit: Be able to increase economic returns, develop resources and promote local economic development to some extent.

In connection with the above four benefits, the most important thing is to estimate potential tourist sources. The higher the popularity of tourism resources is, the larger the number of potential tourists and the more the social and economic benefits will be. On the other hand, the depth of resource survey directly affects the environmental and resource benefits of future development. These two aspects are worth noting.

7.1.3 Survey Emphases

7.1.3.1 Survey of Tourism Resources in Big Cities, Along Traffic Lines and in Densely Populated Areas

Since they serve tourism, tourism resources are of practical significance only if they attract tourists. And the more tourists they attract, the greater their significance will be. Before starting their travel, every tourist would repeatedly weigh whether the time, energy and money they need to spend match the volume and level of tourism resources, just as people would always weigh the relation between price and value when shopping. Therefore, surveyors must consider the psychology of tourists and follow the law of value. For this reason, the survey should be carried out around big cities, peripheral areas of traffic lines and densely populated areas. These areas, owing to their proximity to tourist sources, convenient transportation and a large number of potential travellers, will attract tourists as long as they have distinctive features, notwithstanding their relatively low level and small volume. As for tourism resources far from cities, only if they are extensive, unique and highly graded will they attract tourists with plenty of time, energy and money and be worth development.

7.1.3.2 Survey of Resources in Known Tourist Areas and Their Surroundings

Another focus of resource survey is to carry out in-depth survey of tourism resources in known tourist areas and their surroundings so as to fully tap their potentials. As for these areas, work should be done in the following three aspects:

1. Explore new resources: Fully explore new tourism resources in existing tourist areas and enhance their research degree and sightworthiness; increase the information content of tourism resources so that tourists can learn in sightseeing;

2. Improve supporting functions: Spare no efforts to improve existing supporting functions of tourist areas. A large tourist area must have various functional
   zones. Specifically, some old tourist spots have only tourist attractions but no
   other supporting functional zones, so they fall short of the needs of modern
   tourism. Many of the supporting functional zones hastily established in recent
   years were not planned from a forward-looking perspective, so they fell short
   of the needs as soon as they were completed; attention is not paid to making
   full use of resources and dispersing tourist flows, so that some spots were overcrowded and some spots were ignored; sightseeing was emphasized but rest was
   ignored, so that tourists just walked through without any desire to stay; serious problems relating to water sources and environmental protection also exist
   in many tourist spots. Survey in these areas aims to improve environments,
   enhances supporting functions, upgrades resources and attracts more tourists.
3. Enlarge the scope of tourist areas: Step up surveying the surroundings of tourist
   areas to enlarge the scope of tourist areas, increase tourist capacity, prolong the
   staying time of tourists, provide conditions for supporting functions and form
   integrated new, large and content-rich tourist areas with old tourist spots.

7.1.3.3 Survey of Key New Tourist Areas

Survey of tourism resources requires strategic, forward-looking insight. Tourism
resources which are far from big cities and traffic lines and are sparsely populated
but have high development value should be developed and evaluated in advance in
order to help relevant authorities make decisions.

1. Large featured landscapes: In unfrequented places, there are many little-known
   unique landforms, the discovery of which often leads to the establishment of
   tourist areas, the construction of roads and the development of transportation,
   ultimately forming new economic development zones centring on tourism.
   Examples are Zhangjiajie in Hunan, Nine-village Valley in Sichuan and Five Big
   Linking Lakes in Heilongjiang.
2. Landscapes with special functions: Tourism is sure to develop more deeply and
   widely, with the original pure sightseeing to be replaced by travels in various forms.
   This will boost the exploration of new resources which can serve various tourism
   needs, such as mountaineering, expedition, exploration and hunting. Sport tourism emerging in recent years is based on local resources. For example, skiing, cliff
   climbing, rafting and expedition all require corresponding natural tourism resources.
3. Landscapes suitable for scientific explorations: As new branches of tourism,
   specialized tourism and scientific tourism tend to thrive recently. Despite a
   small number of tourists, specialized tourism and scientific tourism are time-
   consuming, expensive, and yet meaningful. To satisfy the needs of these tourists, it is necessary to explore landscapes of scientific significance, such as
   geologic standard sections, important palaeontological fossil sites, typical mineral deposits, typical or standard geomorphic landscapes, as well as special but
   centralized geology–geomorphy combinations. Glaciers, loess, special animals,

plants, climatic phenomena and peculiar natural phenomena are all exploitable resources as long as they are well worth appreciation and have sufficient supporting conditions. As to such resources, more intensive researches should be made on them to facilitate scientific explorations.

4. Special tourism resources: One of the most important motivations of tourists is to hunt for novelty. Therefore, we must make efforts to survey tourism resources unique to our country, with native characteristics or taking a leading position in the world. Examples are Qinghai–Tibet Plateau landscapes, and pandas and white-flag dolphins in their ecological environments.

5. Survey of cultural landscapes: Cultural landscapes and natural landscapes often complement each other, enriching tourism contents and attracting tourists at various levels. In any survey of tourism resources, a systematic survey should be made on cultural landscapes.

7.1.4 Survey Contents

Tourism resources must have four foundations and three important conditions. Four foundations are given as: rare and peculiar; a certain number and scale; subtle and peculiar combinations of landscapes, climatic phenomena, organisms and cultural landscapes; and high sightworthiness. Three important conditions are given as: important positions; moderate distances; and convenient transportation and high accessibility. To find out the status quo of tourism resources, it is necessary to survey the following factors:

1. Rocks, strata and tectonics of mountains;
2. Mountains, cleuches and caves that constitute landforms;
3. Springs, brooks, waterfalls and lakes that make up waterscapes;
4. Characteristic plants and animals;
5. Meteorological, climatological and environmental factors that positively and negatively affect tourism;
6. Various cultural landscapes that are worth seeing;
7. Sightworthiness or aesthetic characteristics formed by integration of all factors;
8. Locational factors such as position, distance and transportation;
9. Survey and analysis of tourist sources;
10. Positive or negative effects of neighbouring resources on tourist sources;
11. Local economic conditions, reception conditions and dependence on cities;
12. Quantitative survey and qualitative evaluation should be made on resources.

7.1.5 Survey Phases

Survey of tourism resources is aimed to provide basic data for development and construction. Since the degree of future development varies with the different levels and scales of resources, basic data may be of different grades. Survey may generally be divided into three stages:

7.1.5.1 General Survey

This means the check of general conditions based on the data collected, marking relevant scenic spots on small-scale maps (1:100,000 or 1:200,000) and designating prospective areas on the maps. The task of general survey is to investigate and verify known spots or to validate the forecasts made on the basis of other existing professional data. This general survey may be made in a wide range so as to ascertain the distribution and the law of distribution of resources; it may also be made on the status quo of designated areas in a small range in order to ascertain the background of tourism resources and provide basis and advice for follow-up work.

7.1.5.2 Preliminary Survey

Preliminary survey is made systematically on the status quo and it is an in-depth survey of the prospective areas under general survey or prediction. This kind of survey is targeted at the scale, quality, aesthetics and tourist sources of tourism resources in a given area and has results marked on a set of medium-scale maps (1:50,000). It is both an intermediate phase for the exploration of large resources and the final phase for the exploration of medium and small resources.

7.1.5.3 Detailed Survey

Detailed survey is investigative and needs to propose planning ideas. On the basis of preliminary survey, detailed survey should further identify priorities, narrow down the range and have a large-scale map (1:5,000, 1:10,000) drawn. We should pay attention to the collection of data in survey, conduct a thematic study and identification on key problems and sections, make a systematic survey of external conditions needed for future development, as well as put forward planning suggestions for crucial issues.

7.1.6 Professional Survey

One of the important branches of tourism is specialized tourism. Exploitable specialized tourism resources should have not only features of common tourism resources but also professional value. Therefore, in the survey of resources, we should first ascertain their professional value, based on which an in-depth study should be made on professional contents. Only if the study is highly professional can the resources serve professional explorations. For example, the middle and upper Proterozoic strata in Jixian County are thick, clearly stratified, well preserved and highly representative and are of global significance and worth professional explorations. However, to organize formal professional tourism explorations, we also have to make an in-depth research, prepare detailed maps and data, and properly arrange board and lodging and provide convenient transportation. Only when all these requirements are met can we receive Chinese and foreign tourists planning to conduct scientific tourism activities themed on exploration.

7.2 Objectives and Contents of Evaluation of Tourism Resources

Evaluation of tourism resources is conducted on the basis of survey. It includes individual and comprehensive evaluations of tourism resources. Every evaluation requires scientific analysis and feasibility study of the scale, quality, as well as development prospects and conditions of regional resources so as to provide a scientific basis for the development, planning and management of tourist areas. Therefore, the evaluation of tourism resources is an advanced work which directly affects the degree of development and utilization of regional resources and the prospects of tourist sites. In particular, an appropriate evaluation is an important aspect of comprehensive development of tourist regions.

7.2.1 Objectives of Evaluation of Tourism Resources

1. Provide scientific certification for the development of new tourist areas and a basis for the transformation and expansion of fully and partly developed old tourist areas through the evaluation of quality, structure, functions and nature of tourism resources.
2. Provide a series of materials and standards of judgement for the hierarchical planning and management of the country and regions through the identification of the scale and level of tourism resources.
3. Provide experience for rational use of resources and for achieving overall effects and macro effects and prepare for determining the construction order of different tourist sites through a comprehensive evaluation of regional tourism resources.

7.2.2 Contents of Evaluation of Tourism Resources

7.2.2.1 Serial Factors of Tourism Resources

1. Density of tourism resources: Density of tourism resources, also known as abundance of tourism resources, refers to the concentration degree of tourism resources in a certain region. Such a density of resources is one of the important indicators of the development scale, abundance and feasibility of tourism

resources in a region and is also a basic scientific basis for the development and construction of a tourist site. There must be a unified standard for the calculation of density of tourism resources so as to compare the density of tourism resources of different types and scales.

2. Capacity of tourism resources: Tourism capacity, also known as tourism bearing capacity or tourism saturation, refers to the capacity of tourist activities within a certain time and space. In other words, tourism capacity means the activity level of tourists that the characteristics and spatial scale of landscape resources are capable of when the requirements for the protection of the environmental quality of scenic areas are met and tourists' minimum sightseeing needs (psychological induction atmosphere) are satisfied too. It is usually measured by capacity of people and capacity of time. The former refers to the number of tourists per unit area that a tourist spot can hold and it is the basis for designing the land use, facilities and investment scale of a scenic area; the latter means the basic time necessary for touring a scenic area and it reflects the touring routes, contents, sceneries, layout and construction time and other contents of a scenic area. The more complex, implicative and interesting the tourism resources are, the bigger the capacity of time will be, and vice versa.

3. Particularity of tourism resources: This is also known as individuality of tourism resources; it refers to the features of tourism resources. Features are an important factor to evaluate a region's attractiveness for travellers and also one of the preconditions for deciding on the feasibility and advancing of the development of tourism resources. In this sense, the features of tourism resources are the lifeline of tourism development in a region and also an internal force of regional resource effects. In particular, the tourism resources which are unique to a region or rare elsewhere usually constitute the singular attraction sources of the region, so great attention must be paid to the evaluation of particularity of tourism resources.

4. Values and functions of tourism resources: Values of tourism resources mainly include artistic appreciation value, cultural value, scientific value, economic value and aesthetic value; functions of tourism resources usually correspond to their values. The tourism functions of tourism resources with high artistic and aesthetic values are mainly reflected in sightseeing; the functions of tourism resources with high cultural and scientific values are mainly scientific explorations and protection of historical and cultural heritages; the functions of tourism resources also include entertainment, rest, fitness, medical care and business. These values and functions of tourism resources are important indicators of the development scale, degree and prospect of a tourist site, and therefore must be evaluated scientifically and faithfully. Particularly, a rigorous evaluation must be made on their status at home and in the world.

5. Regional combination characteristics of tourism resources. The layout and combination of tourist attractions of different types demonstrate the advantages and features of resources in tourist sites and should be given due attention in evaluation. An ideal combination of tourism resources in a scenic area should feature high density, short mutual distances, coordination of various tourism resources, as well as linear, closed-loop or horseshoe-shaped touring routes.

6. Nature of tourism resources: Any scenic resource has its own particular nature, which must be ascertained and set out explicitly in evaluation, because the nature of a tourism resource determines the utilization function and development direction of the said resource and also has certain influence on the development scale and degree and tourism facilities of a region.

7.2.2.2 Conditions of Developing and Utilizing Tourism Resources Corresponding Conclusions Must be Made on the Following Conditions in Evaluation

1. Locational conditions: They refer to the geographic position and traffic condition of the location of tourism resources. Geographic position is one of the major factors which determine the development scale, route selection and utilization direction of tourism resources, and it has influences not only on the types and features of sceneries but also on tourist sources of tourism markets. No matter how beautiful the scenery is in a tourist area, it is still difficult for the area to attract travellers if the traffic and routes are inconvenient. Therefore, location and traffic condition are the primary factors in evaluating the development of tourism resources.
The key to evaluating the geographical position of tourism resources is its attractiveness. The attractiveness of tourism resources in many tourist areas and spots in the world is enhanced just because of their unique geographical positions. Examples are the following hot tourist sites attracting tourists throughout the world: the Royal Observatory on the south bank of the Thames of London, UK, which is internationally recognized as the starting point of terrestrial longitude and universal time; Garagalli Town, 22 km north of Quito, capital of Ecuador, which is at the equator of the earth; Basel, the third largest city of Switzerland, which is located at the corner of three countries (inclusive of Germany and France); Aachen (adjacent to Belgium and the Netherlands) of Germany; Bratislava (next to Hungary and Austria), the biggest commercial port of Czechoslovakia; Whiyavo of Fiji, which is the easternmost and westernmost city in the world; the westernmost capitals in the world—Nukualofa of Tonga and Apia of Western Samoa; Tromso of Norway, which is reputed as "The Gateway to the Arctic"; cities located at the Tropic of Capricorn and the Tropic of Cancer where the sun blazes down the farthest—Sao Paulo of Brazil, Chiayi County of Taiwan and Jiangkou Town of Fengkai County of Guangdong, China; and, Stockholm, capital of Sweden, which is near the Arctic Circle at the beginning of polar day and polar night. The tourism resources and man-made landmark buildings there have all become ideal memorial places where people flood for sightseeing, for instance, the Greenwich Observatory and bronze monument in London, Equatorial Monument in Quito, the rocket-shaped landmark building in Basel, Aachen Cathedral which is a place of interest under the special protection of the UN, the Danube bridge tower in Bratislava and many landmark towers at tropics. In evaluating these tourism resources and tourist sites, priority should be given to their geographical positions.

The locational factors of tourism resources are also reflected in natural environments and their generation positions. With regard to natural environments, the attractiveness of tourism resources can be enhanced by well-known land features that they are adjacent to, such as famous mountains, rivers, islands, bays, lakes, springs, cities and roads (railways, highways, ancient trails, etc). The generation position of a tourism resource refers to the locational condition of its formation. These characteristics can lead to different psychological location notions to people, such as coldness and warmth, rainy place and sunny place, torridity and coolness, inland and seashore, plain and mountain, and city and countryside. All these have great influence on attractiveness to tourists at different levels and their selection of touring routes and are also the important conditions for evaluating tourism resources.

2. Environmental factors: Environments of tourism resources fall into many types such as natural environment, social environment, political environment and investment environment. The environment here refers chiefly to natural environment conditions such as the quality of climate, vegetation and water. Climate is directly related to the development degree, scale and utilization season of tourism resources. For example, in hot and rainy summer, tourism resources in monsoon areas are prone to disastrous damages such as water and soil erosion, fracture of trees and landslide, as well as rottenness and deterioration of cultural relics; in cold and dry winter, they are subject to frost cracking, avalanche, wind erosion and chilling damage. In arid and semi-arid areas, the development of tourism resources is mainly restricted by water. The problem of water supply must be solved to meet the needs of tourists for drinking water, water use of facilities in scenic areas and cultivation of regional sceneries. In the areas with dense population, land shortage and uncoordinated economic structure, plenty of tourism resources tend to be developed and utilized as economic resources. Tourism resources are eventually destroyed in different degrees by reclaiming wasteland and felling trees, plundering firewood, quarrying for building and so on. In industrial zones and cities with developed industries, many tourism resources are polluted by domestic sewage and waste gas, waste water and waste residue of factories, which not only destroy the ecological balance of natural environment but also pose a disease menace to the tourist areas and lead to deterioration in quality of water, atmosphere and soil. Therefore, in evaluating the development scale and level of tourism resources, a comprehensive analysis covering the quality of soil, atmosphere and water must be made on the influence brought by the aforesaid environmental factors. Besides, the evolution status and consequences of tourism environment must be predicted according to the action mechanism and scope, depth and speed of influence of environmental factors. Only through such a detailed quality analysis based on reliable data and "diagnosis" can a scientific evaluation be made on the degree and depth of development and utilization of tourism resources.

3. Tourist source condition: The number of tourists is a major factor for maintaining and improving the tourism resource effects. It is hard to develop and utilize a site without a minimum number of tourists no matter how good its scenic resources are. Thus, investigating tourist sources and their number is one of the basic conditions for evaluating functions of tourism resources. Investigation of tourist source

markets covers: the tourist source market attracted by a certain kind of tourism resources; features of levels of attracted tourist sources; radiation distance and range; reactions of people when appreciating such resources; and changes in season. In a word, all issues of tourist source markets related to the aforesaid problems are important contents of the evaluation of tourism resources. Generally speaking, in evaluation of partly developed tourism resources, emphasis should be put on their attractive values and functions for intensive utilization, namely tourist groups likely to be attracted, degree of attraction and benefits after intensive development. For undeveloped new tourism resources, we should mainly assess their various attractive values and put forward a reasonable assessment report and find out some rules of different tourism resources in attracting target tourists by referring to the existing statistics of similar tourism resources nearby.

The grades of tourism resources usually determine the target markets of different ranges. Tourism resources with international value and significance may dominate international tourist markets; those with national value and significance may dominate national tourist markets; those with regional value and significance may dominate local markets. Examples of the first case are the Forbidden City and the Great Wall in Beijing, Mausoleum of the First Qin Emperor and terracotta warriors inside it in Lintong and Peking Man Site at Zhoukoudian, Beijing; examples of the second case are the White Pagoda at Miaoying Temple in Beijing which ranks the top in the country, White Horse Temple in Luoyang, Pagoda Forest of Mount Songshan, Zhaozhou Bridge in Hebei and Leshan Giant Buddha in Sichuan; examples of the last case are the tourism resources already developed in various regions.

In evaluating the competitive effect of tourist sources of tourism resources, we should make a comprehensive analysis in combination with their landscape characteristics, geographical positions, traffic conditions and so on, because these factors have influence on target tourist markets and scales of tourism resources to different degrees.

4. Regional economic development level: Development of tourism resources in a region entails a solid economic base, because the construction of a tourist site needs funds, materials, manpower, and scientific and technological capabilities. These conditions are closely connected with the economic development level of this region. Funds are mainly used for recreation, maintenance, beautification, security, sanitation, catering, accommodations, roads, sites, management and water supply, electricity and communication and other infrastructures of the tourist site. A considerable part is used for construction projects such as maintenance, protection, decoration and beautification of tourism resources. Raising of the aforesaid funds will definitely depend on the economic development level of the region. When evaluating the development scale of tourism resources, we should not just rely on investments from outside. What is more important is that we should investigate the economic development status of this region, for instance, the local GNI, the general consumption level, the development status of the average income of residents and the income channels of the main economic sectors. As tourism construction takes time, the funds and materials that can be supplied locally should be further guaranteed for a number

of years consecutively according to the general plan. In particular, the material supply capacity is the major condition for the development of a tourist site. A great quantity of materials such as timbers, steel materials, stone materials and concrete are indeed needed in the development and maintenance of tourism resources, but more materials are needed after the tourist site is opened, and abundant foods and drinks as well as plenty of native produce should be provided for tourists. All these depend on the local aggregate supply of materials.

Manpower is an indispensable condition for developing tourism resources. A tourist site must have adequate economic strength to attract local labour force. The number and income level of labour forces are the basic indicators of a tourism labour market; great importance should be attached to their evaluation. The scientific, technological and cultural capabilities of tourism practitioners are the basic preconditions for the development and superior services of a tourist site, but training on tourism practitioners and managers needs money and time. Higher cultural capabilities of a tourist site may minimize the cost of such training.

5. Construction conditions: Development of tourism resources entails some facility sites, which are mainly used for setting up facilities for sightseeing, entertainment, reception and management, such as sightseeing roads, entertainment carriers, hotels and parking lots. These facilities need different geological, topographic, land and water conditions. Since the development of tourism resources is bound up with the degree of difficulty and quality of the above conditions, these conditions should also be covered in the serial evaluations of development conditions. For instance, the geological and topographic conditions must be taken into consideration in the site selection for tourist restaurants; parking lots must be designed in consideration of convenient access, open field, a small earthwork volume for construction and so on; adequate water supply and sewage treatment facilities must be taken into account for catering facilities. To sum up, the key to evaluating the difficulty of development is to weigh economic benefits, whose indicators include quantity and time. The former concerns the relationship between development projects and investment scale, while the latter is about the relationship between development projects and benefiting time. There must be sufficient economic and technological certifications and technical indicators of construction of every project for the construction plan of a tourist site. In conclusion, only a reasonable evaluation of construction conditions can ensure no waste of money and practical benefits from construction.

7.2.2.3 Tourism Development Sequence

After finishing the evaluation of a series of factors of tourism resources and a series of conditions of the development and utilization, a general development sequence should be provided. That is, the development sequence of tourism resources should be determined based on the difficulty in development and the relevancy of different types of tourism resources determined according to the available quantitative indices.

7.3 Evaluation Methods of Tourism Resources

The evaluation of tourism resources has more than 30 years' history in foreign countries (Bao 1989). By the methods adopted by most scholars, the evaluation of tourism resources can be classified into two types: subjective evaluation and objective evaluation. The former refers to the evaluation made by people according to their own impressions after investigation, and generally through the method of qualitative description, hence also called empirical method. Owing to the limitation of personal opinions, this method often appears to be subjective although it is simple; and, the latter can further classified into two types: supply evaluation and supply demand evaluation. The supply evaluation is made following the in-depth and careful investigation and analysis of attraction functions and values of tourism resources, while the supply demand evaluation is made following the investigation on the demand of existing and potential tourists on the basis of the supply evaluation and through combining of supply and demand. Undoubtedly, the supply demand evaluation is of greatest significance for guiding tourism planning and construction. With the quantitative method adopted, the objective supply demand method is superior to the subjective evaluation method which adopts the way of qualitative description. The evaluation of tourism resources, which has not been performed for a long time in China, mainly emphasizes the qualitative description method and lacks the research on quantitative model at present.

Qualitative analysis will gradually transit to quantitative and qualitative combined analyses in the future, in order to facilitate the establishment of a set of index systems which can evaluate different types of tourism resources. This book describes the evaluation methods of tourism resources only from qualitative and quantitative perspectives, respectively.

7.3.1 Qualitative Evaluation Method

At present, qualitative evaluation has had several evaluation systems, such as the "three three six" evaluation method (Lu 1988) adopted by Lu Yunting of Beijing Normal University. "Three three six" represents "three values", "three benefits" and "six conditions", respectively. Three values refer to historical and cultural value, artistic appreciation value (also named aesthetic value) and scientific research value; three benefits refer to economic benefit, social benefit and environmental benefit; six conditions refer to geographical position and transportation conditions of scenic areas, regional combination conditions of sceneries or scenic classes, capacity conditions of tourism resources in scenic areas, tourist source market conditions, tourism development and investment conditions and difficulty in construction. Huang (1986) of Shanghai Academy of Social Sciences has evaluated tourism resources from two perspectives, with evaluation systems as follows:

1. From the perspective of tourism resources themselves, six standards can be adopted:

   (a) Beauty: The sense of beauty of tourism resources;
   (b) Antiquity: Long history of tourism resources;
   (c) Reputation: Well-known things and the things related to famous people;
   (d) Peculiarity: The specific, unique or rare resources;
   (e) Novelty: The novel resources;
   (f) Value: The resources with use value.

2. From the perspective of environment of tourism resources, corresponding evaluation systems are: seasonality; pollution status, namely tourism environmental quality; connection; accessibility; basic structure; social economic environment; and tourism market. These seven items are within the domains of the natural environment, economic environment and market environment and must be evaluated objectively and qualitatively since they have great influence on the development and utilization of tourism resources.

Additionally, the issue of cost evaluation of resources development was put forward by Huang Huishi. He held that when evaluating tourism resources, we should roughly evaluate unit cost, opportunity cost, shadow cost, social orientation cost, etc.

Because of different description objects, the standards used for the qualitative evaluation vary greatly. Evaluators may create standards according to the actual situation of resources. However, such a method of evaluation must observe three recognized basic principles:

The principle of seeking truth from facts The actual and scientific evaluation should be made on the value and functions of tourism resources, which should not be exaggerated or undervalued.

The principle of sweeping generalization Qualitative evaluation is based on massive first-hand information; therefore, comments should not be too minute and trivial but should definitely and briefly summarize the value, characteristics and functions of tourism resources.

The principle of seeking data Although different from quantitative evaluation, qualitative evaluation should be less subjective to the greatest extent and incline to quantitative or semi-quantitative evaluations, so as to achieve digitization.

7.3.2 Quantitative Evaluation Method

Quantitative evaluation can be classified into two types: individual evaluation and comprehensive evaluation. The former is made with landscape elements as units, such as evaluation of water landscapes, evaluation of meteorological and climatological landscapes as well as evaluation of geological landscapes; and, the latter refers to the overall evaluation of a tourist site or area with all landscape elements integrated. At present, the two evaluation methods are both adopted in China. Hereunder we will describe quantitative evaluation methods adopted by several scholars.

7.3.2.1   Landscape Unit Evaluation Method

1. Research on quantitative evaluation of recreational climate. Climate is not only a landscape element, but also an important influence on tourism, recreation as well as recuperation of people. Many scholars all around the world have been making an evaluation of recreational climate. In order to express the obvious sensory comprehensive effects of meteorological elements, the comprehensive effect of dry and wet bulb temperatures as well as wind speed is used by C.P. Yaglou and W.E. Miller to represent felt air temperature, viz, effective temperature (ET). It can be calculated by diagram or approximate formula ET $= 0.4$ $(T_d + T_w) + 15$ (T_d and T_w are, respectively, dry and wet bulb temperatures in the formula). It is also familiar that temperature humidity index (THI) is used to represent the stifling degree, with the formula as follows:

$$THI = 0.4(T_d + T_w) + 15$$
$$\text{or } THI = T_d - 0.55(1 - RH)(T_d - 58)$$

RH in the equality is relative humidity.

Upon the research on the accelerating effect of wind on heat exchange between human skin and ambient air, P.A. Siple and C.E. Passel put forward the numeration of wind chill index (K_0):

$$K_0 = \left(\sqrt{100V} + 10.45 - V\right)(33 - T_a)$$

Not taking into account the evaporation of human skin and under the condition of full shade, K_0 is the total cooling rate of the atmosphere of 4,186.8 J/m^2 h; V represents wind speed (m/s); T_a represents temperature (°C).

Based on the research findings of predecessor, W.H. Terjung put forward two bio-meteorological indices: comfort index and wind effect index for the bio-climatological classification of the climate all around the world. By the body sense of the majority of people to temperature, relative humidity and wind speed, he classified the former into 11 types: extremely cold, very very cold, very cold, cold, slightly cold, cool, comfortable, warm, hot, sultry and extremely hot; and classified the latter into 12 types: wind frosting exposed skin [$\leq$−1400 (4186.8 J/m^2 h)], extremely cold wind [from −1200 to −1400 (4186.8 J/m^2 h)], very cold wind [from −1000 to −1200 (4186.8 J/m^2 h)], cold wind [from −800 to −1000 (4186.8 J/m^2 h)], slightly cold wind [from −600 to −800 (4186.8 J/m^2 h)], cool wind [from −300 to −600 (4186.8 J/m^2 h)], comfortable wind [from −200 to −300 (4186.8 J/m^2 h)], warm wind [from −50 to −200 (4186.8 J/m^2 h)], wind with unobvious cutaneous sense [from +80 to −50 (4186.8 J/m^2 h)], wind with the cutaneous sense of hotness [from +160 to +80 (4186.8 J/m^2 h), the temperature ranging from 30 to 32.7 °C], wind uncomfortable cutaneous sense [from +160 to +80 (4186.8 J/m^2 h), and the temperature $\geq$32.8 °C] and wind with very uncomfortable cutaneous sense [more than +160 (4186.8 J/m^2 h), and the temperature $\geq$35.6 °C].

The Chinese scholar Liu Jihan has analysed the recreational climate of Mount Huangshan, Mount Lushan, Mount Hengshan, Mount Songshan and Mount Taishan based on the meteorological data of the aforesaid five famous mountains and their corresponding mountain feet of Tunxi, Jiujiang, Hengyang, Linru and Tai'an weather stations during 1976–1980. The analysis data indicate that the comfort index and wind effect index of each mountain in each month are lower than those of the corresponding mountain foot, with the spread the largest in summer, second in late spring and early autumn, and the smallest in early spring and mid-autumn. It is much cooler on the mountains than at the feet of the mountains from late spring to early autumn and is colder in cold seasons, but the climate differentiates slightly in early spring and mid-autumn. Such a climatological difference from late spring to early autumn shows the advantage of mountain climate compared with plain and hill climate and leads to the recreational climate of mountains. The aforesaid quantitative research findings about the recreational climate of the five famous mountains are consistent with the actual feelings of the majority of tourists (Liu 1988).

2. The study on evaluation models of scenic landforms The individual evaluations of scenic landforms have been increasing in China. The tourism evaluation of Danxia landforms by Chen Chuankang of Peking University, Huang Jin of Sun Yat-Sen University, as well as the tourism evaluation of cave landforms of Nanjing University and Institute of Geology, Chinese Academy of Sciences, all are effective trial researches. Here, we will take for example the karst cave studied by Chen Shicai of Institute of Geology, Chinese Academy of Sciences to describe the fuzzy mathematical model of grottoes.[1]

Table 7.1 lists out five major factors, grades and subjections of karst caves. The degree of membership of the length of cave, a secondary factor among explicit numeric factors, is determined according to the following formulas:

$$r_i = \begin{cases} \frac{1}{5}(X - 10) & 10 < X < 15 \\ 1 & X \geq 15 \\ 0 & X \geq 10 \end{cases}$$

$$r_{ii} = \begin{cases} -\frac{1}{5}(X - 15) & 10 < X \leq 15 \\ \frac{1}{8}(X - 2) & 2 \leq X \leq 10 \\ 0 & X \leq 2,\ X \geq 15 \end{cases}$$

$$r_{iii} = \begin{cases} -\frac{1}{8}(X - 10) & 2 \leq X \leq 10 \\ \frac{1}{1.5}(0.5 - X) & 0.5 < X < 2 \\ 0 & X \leq 0.5,\ X \geq 10 \end{cases}$$

$$r_{iv} = \begin{cases} 1 & X \leq 0.5 \\ -\frac{1}{1.5}(X - 2) & 0.5 < X < 2 \\ 0 & X \geq 2 \end{cases}$$

---

[1] Chen (1987).

Table 7.1 Comprehensive table grade subjections of factors of karst caves

Factors		I	II	III	IV
Scale	Length (km)	>10	10–2	2–0.5	<0.5
	Area (m²)	>40,000	40,000–10,000	10,000–1,000	<1000
	Depth (m)	>100	100–30	30–0	0
Sediment	Scientific aesthetics	Many kinds, a large area, good forms, rich colours, or a point extremely outstanding	Less kinds, a relatively large area, decent forms, different colours	Several kinds, not too large area, common forms, two or three colours	Single kind, a small area, bad forms, single colour
Arduousness and peculiarity	Steepness, narrowness and water (Qt/s)	High scarps, the vertical drop > 100 m, large-flow underground rivers, $Q > 1$, master cave becoming narrow, long and tortuous suddenly	Relatively high scarp, the vertical drop of 100–50 m, $Q = 1 – 0.6$ for underground rivers	Low scarp, the vertical drop of 0–50 m, $Q = 0.5 – 0.1$	Broadly flat, no scarp, $Q < 0.1$
Cultural history	Prehistoric archeor-ganisms, prehistoric historical sites of ancient human beings (murals, inscriptions and living sites)	Cosmopolitan influence, such as Peking Man Cave at Zhoukoudian, Beijing	Domestic influence, such as Sanyou Cave in Yichang	With historical records and a certain cultural history	Basically no historical culture, or discovered by modern people, with inscriptions of modern celebrities
Transportation, distance from the central city	Transportation means, distance from the central city	Located in a central city, accessible by bus	One-day trip from the central city	Two-day trip from the central city	One-day trip from a county

In particular, X represents the length of cave, r_{i-iv} refers to the degree of membership of each grade and ranges from 0 to 1. The degree of membership of inexplicit numeric factors has also been evaluated by the panel of experts. Fill the grade of membership of all factors in the following fuzzy relation matrix:

$$R = \left\{ \begin{array}{l} r_{11}\ r_{11}\quad\ \cdots \\ r_{21}\ r_{22}\ \cdots\ r_{2m} \\ \cdots\ \cdots\ \cdots\ \cdots \end{array} \right\}$$

$$r_{n1}\ r_{n2}\ \cdots\ r_{nm}$$

In particular, for the degree of membership, $r_{ij} = \mu \mathbf{R}(n_i,\ Vj)$ and $0 < r_{ij} < 1$.
In view of not excessively highlighting the function of weights artificially, the following formula is taken to state weights:

$$a_i = \frac{\sum\limits_{j=1}^{4} r_{ij} \cdot f_i}{\sum\limits_{j=1}^{4}\sum\limits_{i=1}^{5} r_{ij}}$$

$$a_j = \frac{\sum\limits_{i=1}^{5} r_{ij} \cdot f_i}{\sum\limits_{i=1}^{5}\sum\limits_{j=1}^{4} r_{ij}}$$

In particular, $a_i \in (0, 1)$, $\sum_{a_i} = 1$, $f_j = 4, 3, 2, 1$, $f_i = 1$, $a_j = 1$.
Based on the aforesaid vector operations, trial model calculations are repeatedly carried out, and the M $(\cdot, +)$ model is selected finally. In this way, the substitution of weights for evaluation can be avoided, making the degree of membership more reliable and clearly hierarchical. Moreover, the value of membership can be fully considered rather than cut out in the process of operation.

$$B = \text{AOR} = (b_1, b_2, b_3, b_4)$$

The above fuzzy comprehensive calculation formulas are modelled as:

$$b_j = \sum_{i=1}^{4} a_i \cdot r_{ij}, \quad \text{and} \quad \sum_{i=1}^{n} a_i = 1$$

Through the aforesaid fuzzy mathematical calculations, Chen Shicai has obtained the fuzzy vectors and fuzzy index values for the comprehensive evaluation of caves, and he has also made the fuzzy comprehensive evaluation of 40 open caves in China. Without any doubt, such an evaluation is a creation for individual scenic landform judgement.

3. The study on evaluation model of landscape water. The most representative study on the evaluation method of individual landscape is the "quantitative model of evaluation of mountain lakes in Eastern China" adopted by Yu Kongjian of Beijing Forestry University. He introduced the analytical method

of quantitative theory-I to the field of landscape evaluation, achieving the quantitative forecast of landscape quality (scenic beauty) by the qualitative variables of landscape elements. The said research subject is generally carried out in three steps: ① aesthetic evaluation measurement of lake landscapes; ② qualitative measurement of elements of lake landscapes, and seven factors influencing aesthetic judgement of landscapes of people: mountains, vegetation, astronomical phenomena, lake shores, ornaments on lake surface, waters as well as levels; ③ establishment of the quantitative valuation model lake landscapes by computer. Through the aesthetic evaluation measurement of tested landscape groups, Mr. Yu obtained a scenic beauty scale. Afterwards, taking the scenic beauty measurements as reference variables and seven elements of lake landscapes as explanatory variables (categories), he figured out the scores of all categories and the range of all items by virtue of the linear model of quantitative theory-I, thereby obtaining four evaluation models lake landscapes.

7.3.2.2 The Comprehensive Evaluation Method of Landscapes

1. Fuzzy mathematical method. This method is to establish a tourism decision-making model by fuzzy mathematical theories and methods in order to propose the comprehensive evaluation indices of development value of tourist areas. Take for example, the research findings of scenic areas of Mount Nanshan, Lake of Heaven, Mount Dongshan and Kanas Lake of Mount Altay of Urumqi by Xu (1988) of the Institute of Geographical Research on Drought of Xinjiang University. He quantified various constraints of the above-mentioned scenic areas by adopting fuzzy mathematical theories and methods, and then calculated various comprehensive indices influencing the development value of the scenic areas by virtue of establishing a multi-level fuzzy decision-making mathematical model. Finally, he put forward the decision-making scheme upon optimization and sorting. This method is basically consistent with analytic hierarchy process, and it has its own unique features as well, for Xu Jinfa regarded tourism as a complicated set of factors, all of which are fuzzy. Firstly, he termed the aggregation of three factors: tourism resources, tourism objects and media of tourism the grade-1 factor set, which were further divided into grade-2 factor sets, grade-3 factor sets …Only tourism resources (equivalent to tourism objects and the media of tourism) of grade-1 factor set is taken for example in Table 7.2 and are constrained by grade-2, 3, 4 … factor sets as seen from this table. However, the influence of these factors on tourism development value is fuzzy. On this account, Mr. Xu established a multi-level fuzzy decision-making model. Step 1, give certain weight to the aforesaid different factors by their functions; Step 2, outline the primary matrix according to the different conditions of the tourism areas; Step 3, make synthetic calculations for the fuzzy matrix by computer to figure out the comprehensive evaluation results of all levels one by one; Step 4, synthesize the column vectors of the comprehensive evaluation and the row vectors of the grade in the highest level (grade-1 factor set) to figure out the comprehensive index of tourism

Table 7.2 Evaluation factor set of tourism resources

Functions	Historical culture	1. State level; 2. Province level, Region level; 3. County level
	Artistic appreciation	1. Richness of local colour; 2. Depth of sense of history; 3. Height of artistry
	Scientific expedition	1. Natural sciences; 2. Social sciences; 3. Teaching
Benefits	Economic benefits	1. Increasing income and earning foreign exchanges; 2. Promoting economic development; 3. Expanding employment opportunities
	Social benefits	1. Improving intelligence; 2. Constructing beauty; 3. Promoting cultural and ethical progress 4. Promoting friendly communications, scientific and technological exchanges
	Environmental benefits	1. Beautifying nature; 2. Maintaining ecological balance; 3. Protecting ethnic cultures
Conditions	Geographic location and traffic conditions	1. Location; 2. Traffic conditions
	Regional combination conditions of sights	1. Concentration and accessibility; 2. Harmony natural and cultural landscapes
	Capacity conditions of tourism landscapes	1. Capacity of people (person/m^2); 2. Capacity of time (h/scenic spot)
	Investment conditions	1. National investment; 2. Local investment; 3. Departmental fund-raising; 4. Collective fund-raising; 5. Personal investment
	Construction conditions	1. Input and output; 2. Engineering technical conditions; 3. Supply of infrastructures

development value of the corresponding tourist area (spot), and then sort and make the judgement of the tourist areas according to the comprehensive index. The comprehensive indices of all the tourist areas calculated by this fuzzy decision-making model are given as follows: 0.75 for Lake of Heaven Scenic Area; 0.77 for Mount Nanshan Scenic Area; 0.11 for Mount Dongshan Scenic Area; and 0.54 for Kanas Scenic Area. This model has been recognized by experts as a means of development evaluation of tourism resources and tourist areas.

2. Analytic hierarchy process: The quantitative evaluation of tourism resources through analytic hierarchy process has been studied by lots of people abroad. In China, such a research was initiated by Bao (1988), a graduate student of Peking University. In relation to many evaluation projects of tourism resources difficult to be quantified, he separated the complex problems into several hierarchies and then quantified them by virtue of people's subjective judgments. His method is an objective evaluation method integrating people's subjective judgments and is somewhat similar to the fuzzy decision-making model of Xu Jinfa. Taking Beijing for example, he firstly sketched the evaluation index system of tourism resources,

namely the general target evaluation model tree; secondly, he invited a batch of experts to judge the level by weight in manner of filling a form (1 represents equal importance, 3 represents a little importance, 5 represents importance, 7 represents obvious importance and 9 represents extreme importance) and to make judgements of the relative importance of factors in the same hierarchy compared with a certain factor of the upper hierarchy and make personal suggestions. Afterwards, he sorted out, integrated and tested the surveys, and finally sorted the results by computer. The results are given as: 0.7241 for the weight of tourism resources, 0.1584 for the weight of scenic spot scale and 0.1175 for the weight of tourism conditions. The above items belong to evaluation integrated hierarchy. There are 10 factors in evaluation item hierarchy and 11 factors in evaluation factor hierarchy, which are both ranked by the same analytic method. The 10 factors in the evaluation item hierarchy are ranked as: appreciation characteristics, cultural value, regional combination of scenic spots, scientific value, tourism environmental capacity, transportation and communication, food and drink, quality of personnel, guide services and tourism goods. The factors of evaluation factor hierarchy are sequenced as: pleasure, sense of oddity, integrity, historical culture, recuperation and entertainment, popular science education, religious worship, scientific expedition, convenience, safety and reliability, and expenses. According to the above weight rankings, Bao Jigang scored all factors given the full mark of 100 points, thereby obtaining a parameter table of quantitative evaluation of tourism resources (Table 7.3). In light of the score limits in this parameter table, we can score tourism resources under quantitative evaluation item by item. It can be seen that this evaluation method is an attempt to combine quantitative and qualitative evaluations of tourism resources. Despite containing subjective elements inside, this method is much more objective than purely subjective qualitative description.

3. Integrated scoring method. The advantage of this method lies in that it absorbs the strengths of other different methods and adopts easy and flexible calculations. But subjective assumption is the biggest disadvantage of this method. There are many instances of using the said method to evaluate tourism resources. In China, Wei (1984) of Beijing Institute of Finance and Trade put forward this method earlier. He divided tourism resources into six evaluation items: types of resource constituent elements, individual evaluation of elements, combination of elements, tourist capacity, comparison of cultural resources and difficulty in exploitation. These items can be scored with two methods.

Method 1: The above evaluation items are regarded as equally important and the full mark of each item accounts for 1/6. Each item can be further separated into several factors. According to the satisfaction to an item, the factors of the said item are generally scored by five grades: 100, 80, 60, 40 and 20 points given the full mark of 100 points. Afterwards, sum up the scores of the six items. The higher the total score and the average score are, the greater the resource value will be. This scoring method is called equant system scoring method. The formula is as follows:

$$F\sum_i = \sum_{p=1}^{p} pF_i \ \text{ or } F_i = \sum_{p=1}^{p} F_{pi}/p$$

Table 7.3 Parameter table of quantitative evaluation of tourism resources

Evaluation integrated hierarchy	Score	Evaluation item hierarchy	Score	Evaluation factor hierarchy	Score
Resource value	72	Appreciation characteristics	44	Pleasure	20
				Sense of oddity	12
				Integrity	12
		Scientific value	8	Scientific expedition	3
				Popular science education	5
		Cultural value	20	Historical culture	9
				Religious worship	4
				Recuperation and entertainment	7
Scenic spot scale	16	Regional combination of scenic spots	9		
		Tourism environmental capacity	7		
Tourism conditions	12	Transportation and communication	6	Convenience	3
				Safety and reliability	2
				Expenses	1
		Food and drink	3		
		Tourism goods	1		
		Guide services	1		
		Quality of personnel	1		
Total	100		100		

In the formula, $F\sum$ represents the total of the scores of all items; F represents the grand average score of all items; F_p represents the score of each item; p represents the number of the estimated items; i represents the number of the evaluated tourist sites.

Method 2: Give different score proportions, namely weights, to evaluated items according to their different importance. When scoring, weight the initial scores of evaluated items firstly to calculate the final score of each item, and then sum up the scores of all items. The higher the total score is, the greater the resource value will be. This method is called differential system scoring method. The formula is given as:

$$F\sum_i = \sum_{p=1}^{p} X_p F_{pi}$$

In the formula, X_p represents the score proportion of each item and the other symbols are the same as stated before.

Lu Zi of Hebei Normal University has also proposed an evaluation formula of tourism resources, which is similar to the above formulas. Different from

Wei Xiaoan in respect of evaluation items, he classifies tourism resources into 7 clusters, namely scenic meteorological and climatological landscape, scenic geological and geomorphic landscape, scenic aqueous landscape, scenic zoological and botanical landscape, revolutionary memorial place landscape, revolutionary building landscape and landscape of places of historical interest. And then, further classify each cluster into many grade-2 types, which total 68. Upon such an analysis and based on the principle of statistics, he has proposed an evaluation model of tourism resources, which is in agreement with the formation of the above clusters of tourism resources.

$$Z = \sum_{i=1}^{7} a_i \quad a_i = \frac{\sum_{j=1}^{k_i} b_{ij}}{k_j}$$

$$Z = \frac{\sum_{j=1}^{k_1} b_{ij}}{k_1} + \frac{\sum_{j=1}^{k_2} b_{ij}}{k_2} + \cdots + \frac{\sum_{j=1}^{k_7} b_{7j}}{k_7}$$

In the formula, a_i represents the measure of Class i landscape of tourism resources, ranging from 0 to 10 and with $i = 1, 2, \ldots 7$; b_{ij} represents the measure of Element j in Class i landscape of tourism resources, which ranges from 0 to 10; k_i represents the number of the elements of Class i landscape of tourism resources. Research Office of Beijing Tourism Institute has also designed an evaluation system of tourism resources. The method they adopt is also a scoring method by hierarchy and factor. Its evaluation items include: attractiveness, exploitation conditions and tourism benefits, each of which includes many secondary factors, totalling 19. After the experts scored, they calculated the total score of each item by computer and then sorted out the items.

4. Exponential notation method. This quantitative evaluation method is generally performed in three steps. Step 1, survey and analyse the development and utilization situations, attractiveness and external regional environments of tourism resources. The survey is required to get accurate statistical quantitative data so as to lay the foundation for the quantitative evaluation of tourism resources; Step 2, survey and analyse tourism demands. The survey contents mainly include: demands of tourists, population composition of tourists, length of stay, tourist spending trends, demand structure and demand rhythmicity. Step 3, make the general evaluation, that is establish several quantitative models expressing the relationship among characteristics of tourism resources, tourism demand and tourism resources on the basis of Step 1 and Step 2. Generally, exponential notation method is adopted, with the formula as follows:

$$E = \sum_{i=1}^{n} F_i M_i V_i$$

In the formula, E represents the evaluation index of tourism resources; F_i represents the weight of Resource i in all resources; M_i represents the characteristics and scale index of Resource i; V_i is the index of travellers' demand for i types of resources; and n is the total number of types of resources.

When evaluating tourism resources, Franco F Ferrario of South Africa finalized the total value of tourism resources through combining demand index with the availability (i.e. supply) to tourists. He calls tourism potential attractiveness of scenic spots tourism potential index. The calculation formula is given as:

$$I = \frac{A + B}{2}$$

In the formula, I represents tourism potential index; A represents tourism demand; B represents tourism availability, namely supply.

The index can indicate the actual usability of a scenic spot, which fully demonstrates the tourism attractiveness of the said scenic spot no matter what its characteristics are. But in order to calculate the potential index, it is necessary to meaningfully quantify the above two variables (demand and supply) of every scenic spot. Since the quantification of demand index has been stated above, the remaining problem is how to quantify supply. Based on general feelings, observations and experience of people, Ferrario has selected six criteria: seasonality, accessibility, authorization, importance, vulnerability and popularity which reflect the basic characteristics of tourism resources, for the evaluation of tourism supply. Then, he let more scholars make subjective evaluations of the six criteria, determined the relative contributions of these criteria by figure upon comparison and ranked them in descending order.

Ai Wanyu, a scholar of China, has proposed a bidirectional index evaluation method of landscape quality, with the formula of landscape quality index (j_z) as follows:

$$j_z = \sum_{i=1}^{n} C_i I_i = C_1 I_1 + C_2 I_2 + C_3 I_3 + \cdots + C n I_n$$

In the formula, n represents the number of evaluation factors, C_i represents weight coefficient (equal to 1), I_1 represents aesthetics, I_2 represents specificity, I_3 represents scientificity and I_4 represents scale.

The calculation formula of landscape protection index (j_B) Mr. Ai adopts is

$$j_B = \sum_{i=1}^{n} C_i X_i = C_1 X_1 + C_2 X_2 + C_3 X_3 + \cdots + C_n X_n$$

In the formula, X_1 represents stability of landscape and its surrounding rocks, X_2 represents destructiveness of structure, weathering and erosion, X_3 represents the particularity of ground foundation and X_4 represents artificial damage. The specific evaluation method is the same as the above hundred mark system scoring method and differential system scoring method.

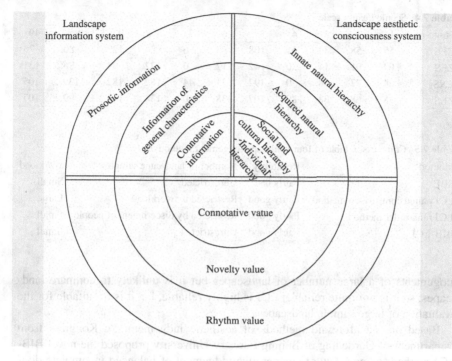

Fig. 7.1 Diagram of landscape aesthetic value system

5. Aesthetic evaluation method. Aesthetic evaluation of landscapes is a complex system of aesthetic consciousness structure, which is under the control of four factors, namely geographic differentiation of population and its surrounding landscapes, differences in acquired aesthetic quality, impacts of social and cultural factors and individual psychological characteristics of people. These hierarchies are in parallel with landscape information system's three hierarchies: prosodic information, information of general characteristics and connotative information, thereby generating three hierarchies of landscape aesthetic value system (Fig. 7.1).

Nowadays, many methods are adopted to make a landscape aesthetic evaluation. Scoring method—scenic beauty estimation (SBE) and method of comparative judgement—law of comparative judgement (LCJ) is generally recognized as the best methods abroad (Yu 1988). Both are evolved from the method of attitude measurement of Thurston and Totorgerson, but they differentiate mainly in specific measurement procedures. SBE mainly takes slides as the media for scoring item by item, according to which a scenic beauty scale is made to reflect the beauty degree of varied landscapes; LCJ has two ways of comparison and judgement: paired comparison method and rank-order method. In short, the above two methods have advantages and disadvantages. SBE is advantageous in making aesthetic

Table 7.4 Scenic beauty scale

Photo no.	1	2	3	4	5	6	7	8	9	10...	49
ZPE	55	58	33	61	108	3	6	86	122	80...	76
ZES	63	91	14	30	142	17	0	122	94	86...	131
ZNS	64	77	58	0	192	31	44	182	182	139...	105
ZLS	38	68	38	89	107	38	60	138	146	99...	107

Table 7.5 Comparison table of four landscape judgement methods

Method	Reliability	Number of landscape samples	Workload
SBE	Fairly bad	Unrestricted	Small
LCJ paired comparison method	Fairly good	Restricted by workload	Large
LCJ rank-order method	Fairly good	Restricted by discernment of people	Small
BIB-LCJ	Fairly good	Unrestricted	Small

judgements of a large number of landscapes but it is unlikely to compare landscapes, so it is not quite reliable; LCJ is highly reliable, but it is unsuitable for the evaluation of large sample landscapes.

Based on the aforesaid methods of aesthetic judgement, Yu Kongjian from Department of Gardening of Beijing Forestry University proposed the novel BIB-LCJ method, namely law of comparative judgement of balanced incomplete divisions, which eliminates the flaws of the two methods by means of the aesthetic evaluation method. He firstly selected 50 photographs as judgement materials and then took four groups: the public, experts, non-professional students and professional students as the judges. According to psychological statistics, he made a scenic beauty scale of landscape aesthetic judgement according to four tested groups as well as a rating scale (partial) (Table 7.4). In the table, ZPS, ZES, ZNS and ZLS, respectively, represent aesthetic measurements got by the results of aesthetic judgements of the four tested groups.

According to the aforesaid landscape scenic beauty, Mr. Yu conducted some related analyses: distribution test of scenic beauty scale; correlation analysis, regression analysis and difference analysis of landscape aesthetic characteristics of different groups of people. These analyses show that BIB-LCJ is currently the best measurement method of landscape aesthetic judgement, which features high reliability, small workload, unlimited number of landscape samples and suitability for the judgement of both large and small sample landscapes (Table 7.5).

Tourism Department of Hangzhou University has also proposed an evaluation model of the beauty of surficial landscapes, with its parameters and levels of score shown in Table 7.6.

Integrated value method The exploitation value of a tourism resource is mainly determined by its integrated value, which is essentially the total of the values of

Table 7.6 Table of parameters and scores of landscape beauty

Parameter	Weight (%)	Scores				
		10–8	8–6	6–4	4–2	2–0
1. Beauty of image	20	Very beautiful	Beautiful	Fairly beautiful	Average	Not beautiful
2. Beauty of colour	10	Very rich	Rich	Fairly rich	Average	Not rich
3. Niceness of sound	10	Very nice	Nice	Fairly nice	Average	No sound
4. Dynamic beauty	10	Very beautiful	Beautiful	Fairly beautiful	Average	No dynamic beauty
5. Beauty of artistic conception	20	Very beautiful	Beautiful	Fairly beautiful	Average	Not beautiful
6. Beauty of combination	10	Very excellent	Excellent	Fairly excellent	Average	No beauty of combination
7. Special effect	10	Very strong	Strong	Fairly strong	Average	No special effect
8. Local customs and myths	10	Very rich	Rich	Fairly rich	A little bit rich	None

Table 7.7 Parameters and scores of the evaluation model of landscape value characteristics

Parameter	Weight (%)	Levels of score				
		10–8	8–6	6–4	4–2	2–0
1. Type of elements	10	Very comprehensive	Fairly comprehensive	Fairly many	Not many	Not comprehensive
2. Beauty	25	Very beautiful	Beautiful	Fairly bad	Average	Not beautiful
3. Peculiarity	15	Very Rare	Rare	Fairly rare	Fairly common	Very common
4. Scale	15	Magnificently large	Very large	Fairly large	Fairly small	Very small
5. Value of history, culture and science	25	Extremely high	Very high	Fairly high	Average	Not high
6. Scenic combination	10	Perfect	Very good	Fairly good	Average	Not good

the above items. Therefore, evaluations are made item by item first, and then sum up the evaluation values of all items. Generally, the evaluation is made by three items. Take the evaluation model proposed by Hangzhou University for example.[2]

1. See Table 7.7 for the evaluation model of landscape value characteristics. In Table 7.7, a submodel may also be made for each evaluation parameter, such as beauty model.

---

[2] The teaching materials of short course for tourism resources exploitation and evaluation

Table 7.8 Parameters and scores of the evaluation model of environmental atmosphere

Parameter	Weight (%)	Levels of score				
		10–8	8–6	6–4	4–2	2–0
1. Environmental capacity	40	Very large	Large	Fairly large	Fairly small	Very small
2. Green coverage ratio:	20	Very high (>90 %)	High (>75 %)	Fairly high (>50 %)	Fairly low (<30 %)	Very low (<10 %)
3. Safety and stability	10	Very good	Good	Fairly good	Fairly bad	Very bad
4. Comfort	20	Perfect	Excellent	Moderate	Fairly bad	Very bad
5. Hygiene and health standard	10	Extremely excellent	Very high	Fairly good	Fairly bad	Very bad

2. See Table 7.8 for the evaluation model of environmental atmosphere. It is worthwhile to mention the evaluation model of environmental quality of tourist areas put forward by Ai (1987) of National Remote Sensing Center of China, Changsha Branch:

$$P_I = \sum_{i=1}^{n} C_i P_i = C_1 P_1 + C_2 P_2 + C_3 P_3 + \cdots + C_n P_n$$

In the equation: P_I stands for index of environmental quality of tourist areas; P_1 stands for air quality; P_2 stands for acid rain; P_3 stands for earth and parent material; P_4 stands for the quantity of water on the earth's surface; and P_5 stands for the quality of ground water. Before evaluation, these factors can be assessed and graded by means of pollution index, weight synthetic pollution index, Nemerow index and double indices and are then submitted to experts for final comprehensive evaluation. According to the index of environmental quality, Mr. Ai divides areas into three grades: first grade: extremely clean grade, unpolluted or single index exceeding background value, $P_1 > 80$ points; second grade: clean grade, single index exceeding standards but not seriously, or single index involving poor quality of natural environment, 60 points $< P_1 < 80$ points; third grade, unclean grade, more than two indices exceeding standards, $P_1 < 60$ points, and major measures urgently needed to be taken to improve environmental quality.

3. Take the model designed by Hangzhou University as an example of models of evaluation of conditions of development and utilization (see Table 7.9).

Table 7.9 Parameters and scores of the evaluation model of conditions of development and utilization

Parameter	Weight (%)	Scoring				
		10–8	8–6	6–4	4–2	2–0
1. Market location	20	Extremely excellent	Excellent	Moderate	Fairly bad	Very bad
2. Industrial economic base	10	Strong	Good	Moderate	Fairly poor	Very poor
3. Accessible traffic conditions	20	Comprehensive traffic network, quick and convenient	Non-stop and quick trains pass, and very convenient	Branch lines pass, single line and transit	Near branch lines, slow and inconvenient traffic	No accessible traffic line
4. Distance to base	15	<20 km	20–60 km	60–100 km	100–200 km	>200 km
5. Infrastructure condition	15	Excellent, well-assorted and enough facilities	Good supporting facilities	Moderate	Fairly poor supporting facilities	Very poor and shortage of facilities
6. Dispersity of scenic spots	20	<2 km	2–10 km	10–50 km	50–100 km	>100 km

Reference indices of industrial economic base: (a) gross industrial and agricultural output value per capita of the tourist areas (RMB/person); and (b) national income per capita of tourist areas (RMB/person)

Table 7.10 Weight parameters of the model of evaluation of conditions of development and utilization

Aspect	Weight (%)
1 Characteristics of landscape value (I)	40–50
2. Environmental atmosphere (Q)	20
3. Conditions of development and utilization (K)	30–40
Total	100

After the aforesaid three evaluations are completed, the comprehensive evaluation coefficient can be calculated in this way: Determine the weight of each aspect and make calculation based on the basic formula of model scoring. Weight parameters put forward by Hangzhou University are shown in Table 7.10. Model evaluation equation is the same as above, namely

$$F\sum_i = \sum_{p=1}^{p} X_p F_{pi}$$

According to the aforesaid weights, the comprehensive evaluation coefficient is calculated as follows:

$$F\sum_i = \frac{1}{10}(4.5I + 2Q + 3.5K)$$

Through calculation, the higher score a scenic area gets in the comprehensive evaluation, the higher development value it will have.

Additionally, the comprehensive evaluation model of scenic areas designed by Chen Shicai is different from that put forward by Hangzhou University and is now provided here for reference only (Table 7.11).

Table 7.11 The indices for comprehensive evaluation of scenic areas

Factors	I	II	III	IV	I	
1. Scale	Capacity (length, width and height), area	Very large	Large	Average	Fairly small	Small
2. Scientificity	The types and quantity of deposits and the level of difficulty of the generation of particular deposits	Great scientific value, large variety, more deposits with higher difficult level	Large, many, generated	Average	Fairly bad	Bad
3. Aesthetic beauty	Shapes, colours and sound of deposits	Very beautiful	Beautiful	Average	Fairly bad	Bad
4. Danger and peculiarity	High, steep, narrow and large, quick-flowing and deep water	Very dangerous	Dangerous	Average	Not quite dangerous	Not dangerous
5. Environmental quality	Gas, water, solid, biology and sunlight	Very good	Good	Average	Fairly bad	Bad
6. Cultural and historical heritage	Ancient life, ancient man, celebrities in the past, modern celebrities	Very good and many	Good or many	Average	Fairly bad	Bad
7. Traffic and distance to central city	Level of traffic convenience and distance to central city	In the central city and bus available	One-day tour to central city possible	Two-day tour to central city possible	One-day tour to county seat possible	Two-day tour to county seat possible
8. Network	Related to many tourism landscapes	More than three tours possible in a day	Two or three tours possible in a day	One tour possible in a day	One tour possible in two days	One tour possible in three days or more

(continued)

Table 7.11 (continued)

Factors	I	II	III	IV	I	
9. Counter-influences after development	Tourism promoting various industries, social effects and exchanges	Very good	Good	Average	Fairly bad	Bad
10. Social and cultural background where the tourism subject exists	Education level and the interest in land-scape cultures	Very good education and great interest in landscape	High and interested	Average	Fairly bad	Bad
11. Others	Other important aspects	Very good	Good	Average	Fairly bad	Bad

References

Ai, W. (1987). Classification and grading of tourism resources. *Tourism Tribune, 3*, 40.

Bao, J. (1988). Preliminary study on quantitative evaluation on tourism resources. *Arid Land Geography, 11*(4), 57–59.

Bao, J. (1989). *Quantitative evaluation of Beijing tourism resources. Beijing tourism geography* (p. 7). Beijing: China Travel & Tourism Press.

Chen, S. (1987). Fuzzy evaluation of tourism earth scientific resources. *A Thesis for Mount Tianmu Academic Seminar on Tourism earth-science.*

Huang, H. (1986). Evaluation on tourism resources. *Collected Essays on Tourism* (1).

Liu, J. (1988). Preliminary analysis of recreation climate in several famous mountains in East China. *Tourism Tribune (Supplement)*, 47.

Lu, Y. (1988). *Modern tourism geography* (pp. 117–130). Nanjing: Jiangsu People's Publishing House.

Wei, X. (1984). Study on comprehensive evaluation on tourism resources. *Tourism, 3*, 11.

Xu, J. (1988). Another tourism scenic star in Xinjiang–Nanshan Scenic area of Urumqi. *Tourism Tribune (Supplement)*, 92–100.

Yu, K. (1988). Quality appraisal of natural landscapes. *Journal of Beijing Forestry University, 10*(2), 1.

Chapter 8
Tourism Development Planning

8.1 Theoretical Bases of Tourism Planning

Although tourism planning is a part of the general strategic planning system of the state, its complexity, hierarchy and scientific involvement are rarely found in other departmental planning, because, as a comprehensive undertaking, tourism have both its economic implications and its cultural contents. Economically, tourism involves industry, agriculture, traffic, trade and commerce; culturally, it involves many disciplines and fields. Therefore, tourism planning is a multi-level matrix planning involving many procedures, departments and disciplines. A series of theoretical and practical issues need to be addressed. These issues involve the nature, objective, characteristic, principle, content, category, step and method of tourism planning.

8.1.1 Essence of Tourism Planning

Tourism planning is a departmental planning in overall planning. It is a conception of the future status of tourism or a comprehensive and long-term plan for tourism development. It is different from regular city planning, which primarily aims to design and deploys city as an economic, cultural, political and social entity, while tourism planning is required to choose the objectives and approaches of tourism development, which cannot be solved in city planning. Obviously, tourism planning has its distinctive contents and requirements.

© Springer-Verlag Berlin Heidelberg and Science Press Ltd. 2015   185
A. Chen et al., *The Principles of Geotourism*, Springer Geography,
DOI 10.1007/978-3-662-46697-1_8

8.1.2 Objectives of Tourism Planning

Planning should start with defining its overall objective and requirement, because any specific objective is determined or arranged under the guidance of the overall goal or objective. Overall objective refers to the general strategic destination and requirement of such planning. Since tourism planning is a part of the overall development strategy of a country or a region and serves the overall strategic objective, its overall objective will be: from the tourism sector or system, to meet the vector and requirement set out in the overall plan of national economy and social development, and develop tourism in a planned and systematic way rather than in a blind and random manner.

Specific objectives are easy to formulate after the overall strategic guideline is determined, because the overall strategic guideline, as the general programme for all social and economic activities in a country in the future, is the basic starting point to determine the specific objectives of the tourism economy department. For example, the seventh Five-Year Plan (1986–1990) specifies that China should aim to grow into a developed tourism country in the world and create an accumulative revenue of several hundred million dollars from tourism by the end of the century. To that end, attention should be paid to the following requirements in determining the development phases and tourism plans of respective tourism areas:

1. Tourism planning should approach from three levels: national, regional and local, which should be coordinated, interrelated and mutually restricted. National tourism planning is based on the demands of domestic and foreign tourists and the status and role of tourism in our national economy; its objective is implemented through regional and local planning. Compared with national tourism planning, regional tourism planning is medium-level planning and local tourism planning is low-level planning. Tourism planning at a lower level is restricted by tourism planning at a higher level. That is to say, low-level planning is mainly intended to meet the objective and requirement set out in national or regional tourism planning. Therefore, it is a basic requirement to appropriately coordinate and balance different levels of tourism planning.

2. Tourism planning also aims to establish a set of structural transformation (linear, nonlinear and topological) modes corresponding to scenic resource structures in order to provide strategic directions and countermeasures for tourism management. Planning is an artificial design system, and its external and internal structures should be in line with each other, namely, not only following the laws of nature but also meeting people's objective and requirement, in order to establish a region-specific tourism adaptive system based on corresponding structural transformation. The tourist reception objective and tourist-oriented supply capacity set out in state-level tourism planning should be in line with the demands of domestic and foreign tourists and the scale and quality of scenic resources owned by the country; and the reception and supply plans proposed in regional and local tourism planning should be in line with local tourist market and scenic resource structure. Apparently, each level of tourism

planning should try to build a corresponding transformation mode suitable for the region in scenic resource structure, tourism behaviour structure and tourist market, which is an important objective to strengthen systematic management of modern tourism.

3. Another objective of tourism planning is to bring into full play regional advantages, fostering strengths and circumventing weaknesses, in a bid to maximally boost tourism economy. As regions differ in location condition, resource quality, tourism environment and service and reception capacity, the development of tourism is inevitably influenced by two conflicting factors: propelling factors and obstructing factors. Tourism planning needs to objectively weigh the effects of these two factors and formulates tourism development targets and countermeasures based on local conditions so as to better balance tourism demand and supply. It is not only a vital measure for bringing into play the integrated effect of regional system and speeding up the development of tourism economy, but also one of the basic objectives which tourism planning should meet.

8.1.3 Characteristics of Tourism Planning

Tourism planning is different from other departmental planning, which is because tourist activities are extensive, hierarchical, comprehensive, rhythmic and capable of driving other activities. Tourism planning should take into account the psychological needs of tourists of different genders, ages, occupations, educational levels and interests; consider the benefits and trends of tourists at global, regional and local levels; make efforts to meet reasonable requirements of various sectors including industry, agriculture, traffic, commerce, cultural relic, environment, religion and gardening; and suit the natural characteristics of the respective regions. Only in this way can tourism planning be regarded as satisfactory and characteristic. Such planning is practicable and can weather various information feedbacks. The main characteristics of tourism planning are given below.

8.1.3.1 Objectivity

Like any other planning, tourism planning is the subjective response to the objective world, or more specifically, the scientific response of people's subjective consciousness to scenic objects. Scenery is objective landscape, and landscape is scenery combined with region, or is a general term for various scenes on the earth's surface; and the earth's surface and regions are objective geographical entities and regional units of tourist activities, including regional units formed by regional variations and regional contacts between comprehensive tourism regional activities and special tourism regional activities. Both comprehensive tourist areas and special tourist areas are objective entities. Tourism planning aims to remove non-scientific subjective factors, endeavour to fully reflect the objective reality of tourist areas, and meet national principles and policies and the development laws of nature and economy. In doing so, what is essential is whether the contents of planning can reflect objective reality and the ability of planners. In that light, tourism planners should collect the internal information about the region or the country, including tourism resources, tourism environments and service and reception facilities, and external information relating to the region or the country, including tourism demands of tourist market, and then based on the two surveys, appropriately handle the interrelationship between tourism demand and supply and make efforts to balance the two. At the meeting held in the capital of Czechoslovakia in 1964, the World Tourism Organization emphasized that tourism planning "should include these two basic surveys", which aims to make the subjective design system agree with the objective system.

8.1.3.2 Regionality

Tourism planning is the spatial layout of tourist activities in a region. Generally, it consists of determining objectives, analysing approaches and making choices. Objectives refer to the objectives in a specific region; approaches and choices are approaches designed and choices made in these approaches to achieve the tourism development objectives of the specific region. Every scenic region has its measurable boundaries and area instead of transregional changeable planning, flexible planning or virtual planning which cannot be substantialized on the map. Therefore, regionality is the basis of tourism planning, and earthscientifically it usually has three features. First, the region features a certain degree of integrity of natural conditions, environment and geographical background (including natural and cultural landscapes), such as basins, plains and some river; second, the development and utilization of regional tourism resources have formed very distinct regional tourist activities such as sightseeing, history, sports, rehabilitation and religion; and third, there are central tourism cities with traffic routes. Tourism centre is not only a place gathering regional and interregional tourism resources and tourists but also the political, economic and cultural centre of the region; traffic route is the necessary means linking tourist market and tourism resources. Only when the aforesaid three features are organically combined can a complete tourist area be formed, and tourism planning is exactly an artificial design system matching such a region.

8.1.3.3 Hierarchy

Tourism planning is hierarchical. Generally, a large tourist area can be divided into first-level area, second-level area and third-level area. If a country is taken as the first-level planning area, then its second level is region and under region is the third level. Examples of the former are Beijing tourist area, north-east tourist area and north-west tourist area; and examples of the latter are downtown tourist area and Mount Xiaoxi tourist area in Beijing, Dalian Seaside tourist area and Mount Changbai tourist area in North-east China, Xi'an tourist area and "Silk Road" tourist area in North-west China, etc. Of such levels of tourist areas divided by range, each higher level is an organical aggregate of several lower levels; after several lower levels have aggregated into a higher level, they take on a new quality and new characteristics. Clearly, these different tourism regional levels are mutually penetrated and restricted, so that their planning is inevitably characterized with criss-cross matrices. The intersection of horizontal and vertical planning is the balance point of planning. Matrix organization method used in tourism planning provides conveniences for the application of mathematical models and electronic computers.

The hierarchy of tourism planning is reflected in the demand and behaviour of tourists as well as in regional system. For example, Professor Chen Chuankang divides the structure of the tourism industry into three levels:

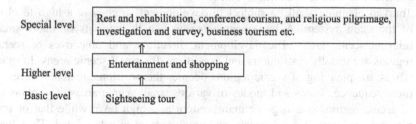

8.1.4 Principles of Tourism Planning

Tourism planning should follow five principles below:

1. Tourism planning shall be in accordance with state regulations and policies on management of scenic areas. Taking China as an example, including *Provisional Regulations on the Protection and Control of Cultural Relics* issued in 1961, *Law of the People's Republic of China on the Protection of Cultural Relics* passed in the Fifth National People's Congress of the People's Republic of China in 1982, *Forest Law, Environmental Protection Law*, and *Provisional Regulations on the Administration of National Parks* issued in 1984, These government laws, regulations and policies are the basic basis for planning of scenic areas, and specify the content and scope of scenic areas, value and level of scenic resources, comparison and certification of various schemes based on opinions of relevant sectors, experts and the public, and land management of scenic areas.

2. Tourism planning is a multi-level professional work involving many procedures and sectors, in which the planners must make careful investigation and research with scientific attitude and methodology. For

example, the increase in tourists in a scenic area is generally proportional to the quantity of tourism service facilities, namely the increase in tourists naturally gives rise to an increase in quantity of tourism service facilities. However, such relationship is restricted by many factors such as traffic conditions and touring methods. If external traffic conditions of scenic areas are improved, so that tourists can come and go more rapidly and conveniently, then the above relationship will not hold or will even be inverse. Moreover, new travel modes will appear with the rapid development of science and technology, and improvement of the living standards and modern facilities. The above example shows that we should take various complicated and changing restrictions into consideration when planning our future procedures and results and make correct predictions based on careful surveys. In general, survey results are the basis for evaluation of plans and prediction. The more information is collected, the more sufficient the basis for planning will be. On the contrary, planning relying on assumptions or preconceived ideas are likely to lead to material errors.

3. Tourism planning shall be subject to development strategy at a high level or of the main system. A large scenic region usually consists of many small relevant scenic areas. The development directions and objectives of scenic regions are usually restrictions and conditions for small scenic areas. In other words, the planning of scenic regions decides the positions, features, development sequence, scales and modes of various small scenic areas. The planning of scenic regions is a large or main system at a high level while that of scenic areas belongs to a subsystem or branch system at a low level. Therefore, planning of scenic areas is preferably preceded by planning of scenic regions. Development targets and scales of scenic areas in large regions shall be analysed and formulated according to restrictions of the large regions so as to make scenic areas at a low level or in subsystems serve the general strategic objectives of scenic regions at a high level or in the main system.

4. In the planning of scenic areas, two or more schemes shall be provided so as to choose the best from them. Article 7 of *Provisional Regulations on the Administration of National Parks* of China stipulates that in the planning of scenic areas it is required to seek opinions from relevant departments, experts and the public and make comparison and certification on many schemes. Such a planning principle is decided by the characteristic of the planning process. This is because the planning process of scenic areas is actually a process of decision-making; in particular, the formulation of planning schemes is "the study on decision-making", which is undertaken by the planners; the choice of final planning scheme is "decision-making action" which is taken by the decision-making leaders or institutions. Therefore, the responsibility of planners is to provide different schemes rather than making choices in place of decision-making leaders or institutions. Both decision-making researchers and leaders should know the characteristic of tourism planning so that in this comprehensive work, it is possible to choose the best scheme through comparison and certification of multiple schemes.

5. Tourism planning shall be flexible within a certain range. Although planning shall follow specific principles, sometimes it should be flexible, which is mainly because the things related to planning belong to the future. "Future" undertakings are infinite, and "future" planning may be a matter of a few years or much longer. It is easy to plan things in the short run, but it is difficult to predict and recognize things which are to happen in the future. As we have limitations in physiology and recognition, the farther the time is, the less attention we will pay. Such limitations are known as "discount" in economics. The farther the time is, the larger the discount will be. Therefore, during tourism planning, it is necessary to consider this discount and possible developments in the future. Changes of people's needs resulting from various social progresses often make it difficult for the planners to accurately predict the future. In that light, the planners should make their tourism planning schemes as flexible as possible so as to accommodate possible changes in the future.

8.2 Contents and Types of Tourism Planning

8.2.1 Contents of Tourism Planning

Article 6 of *Provisional Regulations on the Administration of National Parks* issued by the State Council of China on 7 June 1985 specifies eight planning contents of scenic areas at various levels. Although the planning of scenic resorts is the planning of scenic regions or horizontal planning based on tourist areas, its contents cover the planning of all tourist areas. According to this provision and with reference to experiences accumulated in various regions in recent years, the contents of tourism planning should include the following:

8.2.1.1 Determining the Nature of Tourist Areas

Determining the nature of tourist areas is the primary task and an important content of tourism planning. This is because the nature of a tourist area is the basic factor determining the function and development direction of the tourist area, attracting various tourists and determining the contents and items of tourism facilities. The nature of a tourist area depends on the categories of main tourism resources, and the formation of main tourism resources relies on such factors as geographical location, environmental features, historical conditions and development level of regional economy. In order to distinguish main tourism resources and minor tourism resources, it is necessary to do research in many aspects during the planning. China's tourist areas, except some with single resources, generally have comprehensive touring functions and characteristics. That is to say, most of them are complexes of natural and cultural landscapes with sightseeing, cultural and scientific values. The natural mountains are often tied up with rich cultural relics, historical sites, revolutionary sites, achievements of modern people, poems, legends and myths, which reflect the long history and splendid culture of the Chinese nation and explain why there are so many comprehensive scenic resorts and tourist areas. In that light, it is very difficult to determine the nature of tourist areas. In particular, some large tourist areas are usually associated with economic cooperation zones, provinces, cities and regions which have many landscape resources, which make determination of nature difficult in tourism planning. Additionally, we cannot tag all tourist areas as "comprehensive"; we should distinguish the different importance, values and radiation intensities of the scenes. Only in this way can we identify main tourism resources.

The determination of nature of large tourist areas is often linked with the planning of tourist areas, which will be discussed in detail in the following chapters. With regard to small tourist areas, it is easy to determine their nature. Table 8.1 shows the nature of nine tourist areas in Beijing determined by tourism research team of Beijing CPC Committee and municipal government (Yu and Wang 1989).

China's 44 scenic resorts announced in the first batch have had their nature determined. The nature of most scenic areas was determined based on extensive surveys and opinions of many experts and scholars. For example, in the planning of Five Big Linking Lakes Scenic Area in Heilongjiang Province, we first made an analysis on the four features of this area: unique and magnificent natural landscapes of volcanoes and lakes; natural scenery of expansive northern

Table 8.1 The nature of nine tourist areas in Beijing

Tourist area	Nature
Tourist area in Beijing	Cultural tourist area centred on relics of Imperial City
Tourist area of Mount Xiaoxi and Yongding River	Tourist area of historical interests centred on temples and gardens
Tourist areas in West Beijing	Tourist area of natural landscapes featuring mountains, stones, forests and springs
Tourist area in South-west Beijing	Tourist area of comprehensive landscapes combining places of interests and historical sites centred on ancient ruins, temples, peak forests and caves
Tourist area of Badaling—Ming Tombs	Historical and cultural tourist area centred on famous works and imperial tombs
Mount Yudu tourist area	Tourist area of natural landscapes centred on mountain streams and forests
Huaimi tourist area	Comprehensive tourist area centred on the ancient Great Wall and supplemented by large playgrounds and natural landscapes
Tourist areas in East Beijing	Tourist area of comprehensive landscapes combining forest park, water sports and natural landscapes
Tourist areas in South Beijing	Tourist area of comprehensive landscapes centred on large playgrounds

wilderness; mineral water resource with special medical and economic values; and volcanic landforms and ecosystem of wild animals and plants with high value of scientific expedition. Accordingly, we propose the nature of Five Big Linking Lakes Scenic Area as a comprehensive natural scenic resort which can be used for sightseeing, holiday-making, recuperation and scientific expedition.

8.2.1.2 Delimiting the Scope and Perimetric Protection Zones of Tourist Areas

Tourist areas have their respective scopes and sizes, so it is necessary to make accurate measurement during planning. As some tourist areas serve both as scenic resorts and as nature reserves, certain perimetric protection zones should be delimited. For example, Mount Huangshan is not only an alpine scenic tourist area with granite peak forests but also one of the famous comprehensive nature reserves in China. It measures 154 km² in area, in which 84 km² of five nature reserves are delimited. According to the general strategy of sustained ecological balance and appropriate development and use, a hierarchical protection system was proposed for Five Big Linking Lakes Scenic Area in Heilongjiang Province. The system consists of top-tier reserves such as dragon-like lava, Mount Longmen and Mount Molabu; tier-1 reserves such as Five Big Linking Lakes, Mount Laohei and Mount Huoshao; tier-2 reserves such as Mount Weishan, Mount Jiaodebu, Mount Gelaqiu, Mount Yaoquan, Mount Wohu and Mount Bijia; and tier-3 reserves which are general reserves. External influence areas include water and atmospheric influence areas, whose scopes are all clearly delimited.

Delimiting the scopes of tourist areas or scenic resorts is mainly aimed to facilitate management and specify responsibilities, it is not restricted by administrative zoning so as to keep natural and cultural landscapes complete and meet the tourism needs and facilitate protection and management. In order to protect scenic areas from pollution and maintain ecological balance, some protection zones should be delimited outside the scenic areas. After protection zones are delimited, steles should be set up, boundaries marked and records made.

8.2.1.3 Delimiting Scenic Areas and Other Function Areas

A large tourist area can be divided into several small tourist areas. Different functions should be designed for every area so as to reasonably lay it out. The functions of nine planned tourist areas in Beijing are shown in Table 8.2.

While planning the functional zoning of tourist areas, in addition to determining featured touring areas, it is necessary to mark off living service area, land management areas, recreation areas and parking lots so as to comprehensively arrange tourism facilities. Tourist areas can decide their names on their own. Currently, there are several types of naming in China. The first is to name by "area", for example, Mount Huangshan is divided into six areas: Beihai, Jade

Table 8.2 The functions of nine planned tourist areas in Beijing

Tourist areas in Beijing	It is necessary to meet the rest and recreation needs of non-local tourists and local residents, in particular the requirements of non-local tourists for main tourism resources, forming an international tourism and living base
Tourist area of Mount Xiaoxi and Yongding River	It is a key tourist area of non-local tourists, aiming to mainly meet the needs of non-local tourists for visiting classical gardens and temples, as well as the needs of local residents for rest and recreation
Tourist areas in South-west Beijing	It is an important base to ease Beijing's urban pressure and tourist flow, a channel connecting West Imperial Tombs of the Qing Dynasty and a strategic and preparatory tourist area for developing international tourism, targeting at primarily meeting the rest and recreation needs of local residents, together with international tourism
Tourist areas in West Beijing	It is a major base to ease Beijing's tourist flow, targeting at meeting the needs of local residents for viewing countryside landscapes
Tourist area of Badaling—Ming Tombs	It is an activity base of non-local tourists and tourism-based forex earnings, aiming to explore, expand and improve tourism quality and forex earnings capacity
Mount Haituo tourist area	It is a major base to ease Beijing's urban pressure and tourist flow, targeting at meeting the needs of local residents for viewing countryside landscapes
Huaimi tourist area	It is an important base to ease Beijing's urban pressure and tourist flow, a channel to connect Imperial Summer Resort in Chengde and a strategic and preparatory tourist area for developing international tourism, aiming to enrich tourism contents of local residents and gradually form the competitiveness of international tourism
Tourist areas in East Beijing	It is a potential base of easing Beijing's urban pressure and tourist flow and developing international tourism and a channel connecting East Imperial Tombs of the Qing Dynasty, which targets at meeting the needs of local residents for viewing countryside landscapes
Tourist areas in South Beijing	It is a major base of easing Beijing's urban pressure and tourist flow and a recreation place for foreigners living in Beijing for a long time, which targets at meeting the needs of local residents for viewing countryside landscapes

Screen Tower, Springs, Cloudy Valley Temple, Pine Valley Nunnery and Fishing Bridge Nunnery; the second is by "tourism circle", for example, Hubei Province includes nine tourism circles: Wuhan, Mount Xishan, Mount Mulan, Xianning, Chibi, Jingzhou, Gezhouba, Xiangfan and Tenglong Cave; the third is by "landscape sections", for example, Mount Maoshan Scenic Area in Jiangsu Province was once divided into six sections; the fourth is by "group", for example, Hainan tourist area has three big groups and six small groups.

8.2.1.4   Determining Measures of Protecting, Exploring and Using Resources of Scenic Resorts

Protecting scenic resources and ecological environments should be taken as a strategic task because it is closely related to the inheritance and preservation of historical and cultural heritages of human beings and the future and life of the tourism industry. Many countries in the world treat protecting scenic resources and ecological environment as the lifeline of prosperity of tourism. As put out by Swiss tourism director Mr. Walter, destruction of places of interest and historical sites means loss of the property and environment indispensable to survival of tourism. Therefore, in order to protect and predict tourism resources and environment, it is necessary to evaluate the quality of environment during planning and work out specific measures according to factors destroying the quality of environment. Besides strengthening environmental education and formulating protection laws and regulations, it is also necessary to take protection measures that are well-timed and suit local conditions, e.g. overall protection of ancient cities, restoration of cultural relics and historical sites, supervision on natural ecosystem, treatment of tourism hazards and special maintenance of special tourism resources. Contents of tourism planning cover the following six aspects:

1. Protection items and scope. Including protection of resource entities, environmental quality of scenic areas, characteristics and styles, scenic spots and atmosphere of historical sites.
2. Education and management. Formulating regulations and measures in light of local conditions.
3. Rational layout of various types of land. Strictly observing state regulations concerning protection of scenic areas, preventing setting-up buildings disharmonious with the natural environment and historical features of scenic areas, and deciding the reasonable capacity of tourists received.
4. Restoration and maintenance measures. It is necessary to determine engineering measures for restoration and maintenance of cultural relics and historic sites and put forward plans covering the forms, structures, materials and technologies for restoration of cultural relics and historical sites "as they were".
5. Stressing maintenance of ecosystem balance. Emphasis should be put on planning measures to maintain ecosystem balance of scenic resorts, e.g. specific

approaches to protecting and improving the greening of ecological environment and stopping water from being polluted.

6. Control of "tourism hazards". It is required to remove all possible "tourism hazards" after scenic areas are opened to the public. Examples include mechanical treatment, biological elimination and chemical disinfection of tourism garbage.

All in all, it is necessary to put forward related measures for specific protection objects. For example, in order to protect air quality, traffic volume must be controlled, low-sulphur and low-ash energies should be used and exhaust gas treatment devices should be installed in vehicles. To protect surface water, it is necessary to determine the number of tourists in peak hours and install sewage treatment equipment according to self-purification capacity and supply and demand of surface water. To reduce noise, routes inside and outside scenic areas should be laid out rationally and detailed measures for preventing noise pollution should be stipulated.

Protection and utilization are two sides of the coin in the comprehensive development of tourism resources and should be simultaneously planned. In other words, we should take measures to protect natural ecosystem and cultural relics and historical sites and to develop tourism resources at the same time so as to integrate protection and development.

8.2.1.5 Determining Reception Capacity of Tourism Resources and Organization and Management Measures of Tourism Activities

Tourism capacity or tourism saturation refers to capacity of tourism activities within certain time and space and is generally calculated by person capacity and time capacity in unit time and area. In particular, it is necessary to calculate peak person capacity and best time capacity, because the calculation result plays a decisive role in limiting tourism flow size and protection of cultural relics, historical sites and ecological environment of a scenic area.

While determining reasonable tourism capacity, we should give play to the integrated effect of tourism resources and comprehensively plan the measures for controlling tourism activities. Tourism capacity includes many aspects including (1) resource capacity: referring to tourism activities that the features and space of tourism resources can sustain within a certain time; (2) capacity of ecological environment: referring to tourism activities which the natural environment can sustain; (3) capacity of economic development: referring to tourism activities which the economic and social development level of the tourist area can sustain; (4) regional social capacity: referring to the largest capacity of social tourism activities that can be sustained according to the population composition, religious belief, local customs and living style of the tourist area; (5) inductive environmental capacity: referring to environmental capacity of tourists remaining most interested. All factors affecting the tourism capacity should be taken into account so as to avoid one-sidedness when determining the capacity. In order to control capacity planning, planners should obtain plenty of exact first-hand information through in-depth survey and research, including survey on the annual growth rate of tourists, monthly number of tourists and its change, tourist flow distribution at any place of the scenic area at any time, and feelings and opinions of tourists about sight-seeing effect. This onsite survey information is a scientific basis indispensable to determine proper tourism capacity and a prerequisite for formulating measures to control environment capacity. On the basis of such survey information, planners can work out a scheme of adjusting tourist capacity from macro- and micro-per-spectives (internal space diversion, time diversion and ticket adjustment).

Besides determining environmental capacity of tourist areas, management measures for organizing sightseeing activities should also be proposed in tourism planning. It is necessary to work out a practicable plan for the following aspects: (1) improving transportation and service facilities and sightseeing conditions; (2) strictly keeping to the sightseeing and reception capacity determined in the planning so as to avoid receiving excessive travellers; (3) making full use of the features of scenic resorts to conduct healthy and beneficial sightseeing and cultural entertainment activities and enhance effects of tourism education and knowledge; and (4) maintaining the sightseeing order of scenic areas and guaranteeing safety of travellers and intactness of scenery.

8.2.1.6 Comprehensively Arranging Public Facilities, Service Facilities and Other Facilities

It is necessary to make overall arrangement for facilities to be built in tourist areas and scenic areas, and the layout, locations, construction scale and form of the facilities. The specific contents are as follows:

1. Network structure and layout of service facilities: This planning puts emphasis on guaranteeing accommodation and rest of travellers. It is required to prop-erly balance the concordance of landscapes and service networks and the living requirements of travellers. Large, middle and small service networks should be set to form a complete system of service facilities.
2. Structural layout of road traffic system: This planning pays attention to forming internal and external road traffic networks in tourist areas. Three aspects should be considered: external and internal traffic routes in scenic areas and transpor-tation facilities for tourists. As for external traffic, it is necessary to consider connection with major regional traffic lines and central towns; with regard to internal traffic, repeated routes of scenic spots should be avoided, meanings of scenic spots should be enhanced, and appropriate rest facilities and commer-cial networks should be distributed along routes. In the planning of facilities for distribution and transportation of tourist flows, vehicle structure, location and vehicle capacity of parking lots should be taken into account.
3. Style and layout of buildings of public and service facilities: Buildings as public and service facilities in the scenic areas should suit local conditions and

local landscapes. While being modern with local characteristics, these buildings may be endowed with diversified styles, with due consideration given to local materials and climate. The spatial structure of buildings should be in line with landforms and surrounding environment, and the sizes of buildings should also be arranged flexibly.

8.2.1.7 Estimation of Investment and Returns

While making tourism planning, it is required to propose investment projects and budgets and analyse the expected returns on investment. It is very important because investment planning is a basic condition to develop scenic areas and the material basis to attain planned target. Planners need to put forward several optional development schemes and investment plans if possible and make full and comprehensive technological certifications on the construction conditions of scenic areas so as to accurately and reasonably use the investment.

Scientific analysis of returns is a key to the optimization of investment schemes. It is known to all that there are many factors affecting tourism benefits at different levels. In terms of content, there are economic, social and environmental benefits; from the perspective of time, there are short-term, medium-term and long-term benefits; from the perspective of factors, there are benefits of resource allocation, layout and operation; and from the hierarchical perspective, there are macro-, medium- and micro-benefits. During tourism planning, we should integrate the benefits from the perspectives of content, time, factor and level and analyse the comprehensive analysis.

Investment in developing a scenic resort covers at least four aspects: protection of resources and environment, infrastructure, improvement of reception conditions and external transportation. The funds for these aspects should be secured and their returns should be evaluated and estimated based on capacity and tourist volume.

8.2.1.8 Other Matters to Be Planned

Besides the above-mentioned contents, other planning contents include tourist source markets, land use, water supply and drainage, power supply, energy, communications, regional economic complexes, tourist talent training and tourist goods.

8.2.2 Types of Tourism Planning

Types of tourism planning are very complex and can be divided from various perspectives. Generally speaking, by the planning content and nature, it can be divided into: tourism industry planning, planning of development and utilization of tourism resources, tourism city planning, scenic resort planning, tour route planning, tourist source organizational plannings, planning of tourism facilities construction, tourism goods planning, planning of tourism talents training, etc.; by time, it includes long-term planning and short-term planning; by geographic scope, it falls into national tourism planning, provincial tourism planning, district and county tourism planning, scenic area (spot) tourism planning; by varying degrees in detail of planning contents, it is subdivided into general planning, detailed planning, comprehensive planning, departmental planning, etc. Outlined below are several different types of tourism planning.

8.2.2.1 Tourism Industry Planning

Tourism is an emerging sector of national economy and is regarded by many countries as a tertiary industry and the backbone of their economy. A few countries highly developed in tourism also take it as the fifth industry to highlight their socio-economic status. The function of tourism is to provide various services for tourists, e.g. food, shelter, transportation, travel, purchase, by virtue of tourism resources and facilities. So, tourism industry planning is actually a planning driven by tourism economy, with emphasis on solving and handling the relationship between tourism demand and supply. Therefore, such planning is also regarded as tourism demand, i.e. tourism supply planning.

As the development guideline of tourism economy at different periods, tourism industry planning should not only clearly set out the direction, scale, speed and objectives of tourism development, but also specific measures to achieve such objectives. As such planning is the basis for annual and near-term tour plans, it is therefore of strategic and directive significance for a country or region.

Tourism industry planning can also be divided into general planning, departmental planning, national planning and local planning. General planning is the planning of tourism economic complex involving many industries and focuses on the coordination relationship between tourism development and other related departments and industries from an overall perspective. Departmental planning refers to the planning of one link in tourism economic structure, e.g. tourist source organization, development and protection of tourism resources, design of tour routes, layout of tourism spots and supply of tourism goods.

What is essential to tourism industry planning is the investigation into tourism demand and supply. The former focuses on analysis on tourism market and forecast on number, direction and interests of tourists in future, while the latter emphasizes investigation into features, peculiarity, volume and environmental capacity of tourism resources and existing reception facilities of scenic areas. After that, a balance should be reached between them. If tourists' demand in a tourist area or spot exceeds the capacity of its tour and reception facilities, it is necessary to add facilities for receiving and diverting tourists, or conversely, reduce investment in reception facilities. In addition, tourism industry planning should also be based on a study on the investment environment, certification of the synthetic effects of investment and discussion on the protection of tourism resources and ecological environments.

8.2.2.2 Planning of Development and Utilization of Tourism Resources

Development and utilization of tourism resources are the material bases for sustained development of tourism. With the development of the tourism industry and expansion of the tourist population, it is necessary to develop new tourism resources, give full play to the efficiency of existing tourism resources and at the same time strengthen protection of tourism resources and ecological environments. To that end, it is necessary to work out plans for development and utilization of tourism resources, so as to develop tourism in a purposeful, planned and systematic manner.

Tourism resource development planning mainly includes: (1) overall survey on tourism resources, covering their types, volume, distribution, characteristics and cause of formation; (2) synthetical evaluation on the development values and conditions of tourism resources, analysis on their location factors and accessibility, evaluation on their functions and quality, and regional combination and environmental capacity of scenic spots; (3) certification on economic returns of tourist resource development, forecast on investment payoff period, profits and impact on local economy, and presentation on temporal sequence of tourism resources development; (4) provision of feasibility report for tourism resource development.

8.2.2.3 Tourism City Planning

Tourism city refers to a city which has the function of receiving foreign tourists and can earn a certain amount of foreign exchanges. Such cities: (i) boast abundant tourism resources and hold a particular fascination for tourists; (ii) have sufficient reception and supply capacity and facilities for tourist transportation, accommodation and commodities supply; (iii) have convenient external traffic, with transportation networks of water, land and air supporting facilities; (iv) have well-assorted urban infrastructure and great advantages in the development potential of tourism. Owing to these advantages, tourism city usually becomes the tourist centre of a country or region and plays a decisive role in its economic development. Therefore, tourism city planning is usually a priority in many countries.

Tourism city planning has many unique requirements. For example, inheritance and development of the city's unique features; conformity of industries, transportation, business and all economic and cultural facilities to the characteristics and nature of the city; comparison between all construction projects and tourism investment of the city; the landscaping and beautification of urban environment. All these principles and requirements should be considered in the planning, so as to produce an integrated effect of tourism city and achieve coordination among city layout, land use structure and ecological environment structure.

Tourism city planning pays special attention to the design of axis, gateway scenery and street sculptures of the city. City axis refers to the symbolic linear space that can organically connects all functions of the city and continually and intensively demonstrate the city's characteristics. It is the skeleton of cityscapes. People can not only have a bird's-eye view of it but also directly feel its image and characteristics in the life, e.g. road axis, river axis, historical and cultural axis, urban greening axis and business culture axis. As cities differ in axis scenery, tourism planners should try their best to enrich the inner beauty of city axis according to its characteristics, and other buildings should also remain in harmony with the city.

In addition to city axis scenery, central places of tourism city are also targets that tourists pay close attention to. The so-called city centre is the gathering place of people in the city, with the function of displaying the city's characteristics, e.g. squares surrounded by magnificent buildings, sightseeing centres full of historic sites, parks or crowded downtown areas. Some cities also have city emblems or memorial architectural landscapes. All these appealing architectural landscapes and scenes in the city centre should be carefully maintained and strengthened in the planning.

The gateway sceneries of the city refer to the architectural landscapes at the entrance and exit of the city's trunk roads. They can be magnificent buildings, or small architectural landscapes or sculptures with the city's characteristics. In history, the Chinese people paid great attention to the design of gateway sceneries such as towers, decorated gateways, small bridges, flower beds, stone sculptures and walls, in order to add and highlight city image. Modern city gateway sceneries usually take buildings at the entrance of the city, i.e. railway stations, airports and highways as landmarks, among which quite a few architectural landscapes are not right for the city's nature and are in need of adjustment and alteration in the planning.

Sculpture sceneries of tourism cities are also very important. They have usually been taken as landmarks of the city due to their elegant images, e.g. the Little Mermaid statue in Copenhagen, Goddess of Victory in Buckingham Palace of England, Five-Ram Sculpture in Yuexiu Park of Guangzhou, Phoenix in Yinchuan and Lying Ox in Xingtai City of Hebei Province. The spatial shapes of these sculptures symbolize the city and are great attractions for tourists, who call it "Solidified Music". As such, they should be carefully studied in the planning.

8.2.2.4 Tour Route Planning

Tour routes are designed by tourism authorities for tourists to conduct tourist activities. They and traffic lines jointly form the network connecting all tourism spots or cities. Examples are Jiangnan Canal tour route, "Silk Road" tour route, Qinghai-Tibet Plateau tour route, etc.

The formation of a tour route is constrained by such factors as tourism spots, regional economic development level, traffic conditions, tourism market and travel time. Tourism authorities may design tour routes independently or through inter-regional cooperation according to these factors. What is important is that tour routes should be distinctive and convenient, because tourists select tour routes in consideration of: "short (journey time), long (touring and sightseeing time), many (varieties of scenic spots), few (repeated routes), high (reputation of scenic spots) and low (travelling expenses)". In particular, the features of scenic spots and convenience of the journey are dominant factors in selection of tour routes. Tourism planners should follow these principles in designing series of different routes for tourists at different levels, conditions and with different interests. Examples are one-day trip, two-day trip …several days' trip, by travel time; hiking tour, cycling tour, tour by bus, train, boat and plane, according to vehicles taken by tourists; common sightseeing tour, featured and special tours, by contents and natures of tourist activities.

Special attention should also be paid to investigation and research in the planning of tour routes, e.g. forecast on flow direction and volume of tourist sources, understanding of the reception capacity of all tourist areas and tourism cities, and insight into tourism environmental capacity and transport conditions, so as to divert tourists from "hot spots" and "hot lines" to "unpopular spots" and "unpopular lines", and thus reach an overall balance in them.

8.2.2.5   Tourist Source Organizational Planning

Tourist source plays an important part in tourism development and a dominant role in tourist activities. The key to prosperity of a tourist area lies in the contest for tourist source and flow. Therefore, tourist source organizational planning occupies an important position in tourism industry planning.

Tourist source organizational planning emphasizes on establishing market competition mechanism, mastering the trend of tourism markets and selecting objectives of tourist source markets, so as to establish tourism facilities, organize supply of tourism goods and intensify market competition countermeasures accordingly.

Contents of tourist source organizational planning are given as follows: (1) Determine tourism target markets and specify the sequence of tourist sources to be attracted. A country or region is sure to have its target markets of tourist sources. To accurately grip the trend of such tourist sources, it is necessary to study the tourism market as bridge and tie between tourism supply and demand, so as to timely grasp tourism information and change market strategy mechanism; (2) strengthen tourism promotion to improve the popularity of tourism resources, which is an important measure in tourist source organizational planning; (3) pay attention to seasonal balance of tourist sources and the balance between hot lines and unpopular lines in tourist source organizational planning. Seasonal balance of tourist sources aims to narrow the gap between slack season and peak season due to climatic and seasonal factors. The situation that the off-season of tourist areas is not so low and the peak season is not bad may be achieved by adding appealing tourist activities or adjusting tour prices in slack season. Tourist source balance between unpopular lines and hot lines can also be achieved by opening new lines and spots, improving service quality and reception credit, setting reasonable tour prices and strengthening contacts with travel agencies.

8.2.2.6 Tourism Facilities Planning

Tourism facilities mainly refer to tourism transport, tourist restaurants and hotels and some service facilities for entertainment and physical and cultural activities of tourists. In particular, transport, restaurants and travel agencies are often regarded as the three pillars of tourism and are the material basis for the smooth development of tourism. The significance and contents of tourism facilities planning are examined below by taking tourist restaurants as an example.

Tourist restaurants are an important indicator of the reception capacity of a country and region. The greater the number of tour restaurants and the higher the service quality are, the more developed the country's tourism will be. It is estimated that there are now over one hundred thousand restaurants in the world, 45,000 of which are in the USA, indicating that the USA is highly developed in tourism, followed by Italy, West Germany, Spain, UK, France, etc. In these countries, revenues from tourist restaurants account for a considerable proportion in their economic income. Statistics show that incomes from guest rooms account for 37.4–54.7 % of tourism revenues in the world, and most of these incomes are foreign exchanges. Therefore, one of the important measures for developing tourism and increasing foreign exchange earnings is properly plan tourist restaurants.

The most important principle in planning the layout of tourist restaurants is to intensify investigation into passenger flow, because the adequacy of tourist sources is the main determinant of the scale, quantity and grade of tourist restaurants. The greater the passenger flow is, the greater the demand for restaurants will be, resulting in higher occupancy of guest rooms. To attract tourists, it is necessary to take into account the location conditions in building tourist restaurants. Generally speaking, those restaurants in places with peaceful environment, convenient transportation and close to cities and scenic areas best suit tourists' psychology and needs. In addition, tourist restaurants should also highlight ethnic styles and local features, which can provide tourists an exotic experience and thus increase their appeals to tourists.

8.2.2.7 Planning of Production and Supply of Tourism Goods

Tourism goods refer to articles that tourists buy during their trip, e.g. food, beverage, daily necessities, souvenirs and handicrafts. These articles have the nature of tourism goods because transfer of their use value is based on monetary consumption of tourists. Travel agencies earn foreign exchanges and revenues from tourism via such sale and exchange. Relevant statistics show that revenues from tourism goods in the world account for over 40 % of the whole tourism revenue, of which more than 60 % are from Hong Kong, which is known as "Shopping Paradise". In China, revenues from tourism goods account for only 1/3, indicating that China's production and supply of tourism goods still need to be further improved. The planning of production and supply of tourism goods mainly include the following contents:

1. Survey tourists' demand: Carefully survey tourists' purchases of tourism goods, and design, develop and operate diversified, multi-format, multicoloured and ethnically featured tourism goods, so as to develop tourism markets appealing to tourists.
2. Properly lay out sales outlets: Properly lay out sales outlets of tourism goods, and continually improve the opportunity and convenience for tourists in buying tourism goods.
3. Step-up commodity publicity: Make tourism goods better recognized by and more transparent to tourists.
4. Improve circulation channels: Make efforts to improve circulation channels of tourism goods to reduce the obstacles in sales of goods, and therefore reduce circulation expenses and production costs.
5. Strive to gain in both fame and wealth: Follow the principle "Gain in both fame and wealth", which can not only give play to tourist goods' function of earning foreign exchanges, but also publicize our country's long history, brilliant culture, beautiful mountains and rivers, outstanding techniques and rapid changes through various goods with Chinese characteristics, so as to enhance Sino-foreign cultural and artistic exchanges and friendly communications.

8.3 Method of Tourism Planning

Tourism planning differs in region and type, but shares similarities in method. In recent years, tourism researchers in many countries have conducted all-round studies and discussions on method of tourism planning. For example, some researchers applied Boolean algebra in planning tour routes; some applied Christaller's Central Place Theory in the planning of tourism city; some even applied systematic science theory in guiding the planning and construction of tourist areas. All the above show that the application of different disciplinary theories and methods in exploratory research on tourism planning has become an international trend of tourism development. Take Boolean algebra as an example. Although it is an important achievement of research on rules of human thinking through mathematical approach and has been widely applied in automation technology and computer logic design, it also opens new approaches for research on tour route planning. Tourism scholars can presume several route plans or several approaches of one plan as A, B or C ..., take the predetermined scenic spot as given K, write Boole expression and establish Boolean equations to make operations, and the result obtained is the ideal tour route (Chen 1987).

Applying systematic science in guiding tourism planning and construction is a major concern of the tourist circle at present. The planning method provided in this section is actually an attempt to determine an ideal planning and construction scheme based on quantitative tourism analysis by applying systematic science theory.

Experiences show that successful planning generally requires four factors: (1) feasible objectives; (2) well-educated executives; (3) an evaluation index system; (4) a good feedback system. In fact, the above basic factors, if linked organically, form the whole planning process, because feasible objectives are based on researches and investigations on the status quo and characteristics of tourism resources, the natural and socio-economic conditions of scenic areas, etc. On the basis of substantial first-hand information, strategic development objectives, guidelines and directions, i.e. the plan, should be put forward. After several plans are worked out, it is necessary to compare, evaluate and select from them. Evaluation of plans requires an evaluation index system as the comparison standard in selecting preferred plans. After a plan is selected, it is also necessary to put it into practice to test its feasibility degree. Such a testing procedure is the feedback mechanism, through which defects in the plan are corrected.

Thus, planning is actually a process from "making assumption" to "finding strategy" and "judging sensitivity", in other words, a process of making selections from several different feasibilities in future. In this sense, planning is also a decision-making process. The significance of the process from understanding to planning lies in the emphasis on the relationship between planning steps and ensuring implementation of the planning by following scientific and reasonable procedures. In the past, many people took planning as only a result (making a plan), but neglected the rationality of planning. This is the root cause of mistakes or failure in planning. We view the whole tourism planning process as a system, which is also called as system analysis procedure.

Tourism system analysis procedure is intended to help us use new modes of thinking such as quantitative or mathematical means to describe and quantify concepts, certify objectives and strategies, establish evaluation standard, insist on testing theories with practices and forecast results, so as to make the plans reasonable and practical. The tourism system analysis procedure is divided into the following seven steps.

8.3.1 Finding Out Constraints and Setting Planning Objectives

Learn about the basic constraints of tourism planning tasks. The basic task of the planning in the early days is the preliminary development of planning objectives, which is only the preparatory stage during the whole tourism planning process, but should be as accurate as possible.

1. Understanding the basic constraints of tourism planning tasks: (1) the background and history of tourism planning tasks, i.e. understanding problems and causes of planning tasks and historical evolution; (2) necessity and significance in solving the above problems, i.e. knowledge about the scenic area planning's environmental status (location factors), nature and influence on its surroundings (in a higher and bigger system); (3) the basis and range that the tourism planning depends on, i.e. policies and regulations and scope of territorial rights.
2. Preliminary development of strategic objectives of tourism planning: (1) specify constraints of tourism planning, i.e. first consider the conditions to be met; (2) specify contents of tourism planning, i.e. all questions and details related to the planning, so as to lay foundation for the establishment of evaluation index system; (3) specify the ultimate goal of tourism planning, i.e. present the depth and standard of the result.

8.3.2 Information Survey

Organize ad hoc survey and submit reports on investigation, analysis, and evaluation and forecasts for the future according to tourism planning objectives.

First, survey of nature:

1. *Survey scope*

    (1) Geological conditions such as geology, landform, hydrology, climate, soil, vegetation and historical evolution; (2) resource conditions such as scenery resources (including natural tourism resources and cultural tourism resources), production resources (including industry, mining, agriculture, forestry, planting, farming, livestock products and handicraft industry); (3) ecosystem including atmosphere, water, fauna, flora and people.

    Second, social survey:
    (1) Social conditions, including management system (management institution, actual management effectiveness, etc.), economic level, literacy, habits and customs; (2) technical conditions, including transportation, water, electricity, communication, accommodation, medicine and materials supply.

    Third, market survey:
    (1) Domestic and foreign demand in the near term; (2) characteristics of similar projects in China and their competitiveness; (3) source of construction and management fund.

    Fourth, the survey on rights:
    (1) Relevant state laws, policies and relevant regulations of provincial and municipal authorities; (2) land ownership.

2. Survey report and evaluation forecasts report: In the event that the information is insufficient, forecasts may be carried out on the basis of relevant information, but due attention should be paid to reliability.

8.3.3 Determining Planning Principles and Establishing an Evaluation Index System

1. Determination of planning principles usually refers to planning of guidelines.
2. Establishment of evaluation index system: (1) Divide targets into several determined indexes, i.e. assort details of the planning for further determination; (2) specify the sequence of priority of indexes and the principles for handling conflicts. Divide all indexes into five grades, i.e. extremely important, very important, important, worth consideration and meaningless, and then weight the above indexes.

8.3.4 Preparing Planning Schemes

Various professionals should be invited for the preparation of tourism planning scheme as it involves many disciplines. The collective combination and horizontal linkage of talents is the basic characteristic of scheme preparation.

1. Try to bring forward methods meeting all indexes required by the planning.
2. Select methods and sum up several alternatives according to index evaluation system.
3. Search and seek schemes: Draft out schemes according to planning objectives and review them according to evaluation index system. If any scheme fulfils requirements, follow up; if not, seek other solutions.

8.3.5 Analysis and Evaluation

Make evaluation by using relevant matrix methods according to tourism planning targets, technological and economic feasibility, rationality in layout and limitation of time.

1. Multiply the evaluation value $f_i(x)$ ($i = 1, 2, 3,\ldots, n$) of each scheme obtained according to requirements of correspondent indexes by relevant W_i in light of the importance of indexes, and then use the formula:

$$U(x) = \sum_{i=1}^{n} W_i f_i(x)$$

and matrix form as the evaluation basis for selecting schemes, so the maximal solutions of $u(x)$ are solutions of various target systems.

2. Bring forward remedies to defects in the scheme upon evaluation

This is the end of the decision-making stage of tourism planning.

8.3.6 Scheme Selection

Scheme selection is completed, chiefly in the following ways, by the consignor of the planning task and specialized evaluation institutions:

1. Oral defence for schemes: Planners introduce and defend the schemes to enable the judges better understand the schemes and select from them.
2. Delphi technique: The judges make comments back to back and select the best scheme upon statistics. The advantage of this method is that every judge can freely express their own opinions, avoiding predominance during selection.

After the scheme is determined, it is necessary to bring forward remedies for defects in the scheme, so as to achieve a satisfactory result to the extent possible.

8.3.7 Planning Implementation

1. To guarantee planning implementation, it is generally necessary to establish a sound management system, economic system and time system (1) Management system: ① person in charge of the decision-making institution should be the local chief executive officer. The action time is also associated with regulatory institutions having money and right and supported by law. ② It is necessary to legislate for planning scheme. Legislation is the guarantee of decision-making and action, and the unity of motive and effect. ③ The executive agency will ensure successful implementation of all decision plans of decision-making institutions and are the watchdog of implementation. It transmits any problem found during the implementation to the decision-making institution via the feedback system. (2) Time system represents the time needed from general tourism planning to all stages of the implementation. (3) Economic system refers to fund sources and their allocation in all stages.
2. The implementation of general planning includes the detailed planning, design and completion of various items.

References

Chen, J. (1987). *Study on evaluation and planning of tourism resources, tourism earth-science research and development of tourism resources* (p. 53). Chengdu: Sichuan Science and Technology Press.

Yu, Y., & Wang, L. (1989). *Protection and development of tourism resources in Beijing, Beijing tourism development strategy* (pp. 157–159). Beijing: Beijing Yanshan Press.

Chapter 9
Protection of Tourism Resources and Tourism Environments

9.1 Tourism and Tourism Environments

Tourism environments are the sum of all external conditions of tourist activities, including social and political environment, natural eco-environment, tourism atmosphere environment and tourism resources.

Social and political environment refers to political situation, public security and residents' understanding of tourism, provision of living services, etc. A stable political situation and sound public security can give tourists a sense of security; local residents' correct understanding of tourism can make tourists feel happy, without being deceived and insulted; sufficient material supply and quality services can make tourists get the enjoyment they deserve. In a word, the quality of social and political environment is closely related to the development of tourism. However, it will not be further discussed here because it is basically a sociological issue.

The natural eco-environment in a tourist area is the complex of the landform, air, water, fauna, flora, etc., in the area. Direct tourism objects may also be viewed as an external environment, big environment or prospect environment. Although it is usually not valued by tourism development authorities because it is not considered as direct tourism resources, it actually has a direct bearing on success or failure of tourism there. With continually developing modern large-scale industrial production, expanding cities and increasing population density, the human living environment is getting farther from nature and conditions are getting worse. With the advancement of science and technology and improvement of understanding, an increasing number of people come to realize that the environment provided by themselves, but sustaining damage and pollution also brings harms to their physical and mental health. As such, all people miss and yearn for nature. "Return to nature" becomes a popular slogan. People may take a trip for different purposes, but a pursuit of both mental and physical enjoyment in a good natural environment

© Springer-Verlag Berlin Heidelberg and Science Press Ltd. 2015
A. Chen et al., *The Principles of Geotourism*, Springer Geography,
DOI 10.1007/978-3-662-46697-1_9

should be the important motive of the majority. In this sense, clean water, fresh air, peaceful environment and dense vegetation are themselves tourism resources.

Tourism atmosphere environment refers to the environment reflecting historical, local and ethnical flavour formed by historical and modern development on the basis of clean, beautiful and less polluted natural environment, primarily consisting of harmonious buildings, recreation facilities and activity sites. What we provide to tourists cannot be a single sightseeing or recreation object, because it may reduce or lose its attraction to tourists if it is detached from the atmosphere environment filled with historical, local and ethnical flavour. A traveller's leaving his regular place of residence for another place is essentially a yearning for exotic atmosphere, or in other words, a pursuit of tourism atmosphere environment in another place. As such, attention to the protection of atmosphere environment is crucial to tourism development.

In a word, tourism is an industry based on attraction to tourists and the resulting tourism flow. The attraction to tourists is the comprehensive tourism environment composed of tourism resources, social and political environment, natural eco-environment and tourism atmosphere environment, rather than an independent building or an activity. The quality of tourism environment has a direct bearing on success or failure of tourism in one region. The environmental benefit of tourism has a more direct impact on its social benefits and economic benefits compared with other sectors. Therefore, the protection and improvement of tourism environment is very important. The impact brought by industrial pollution in China is now drawing increasing attention, while the losses of tourism arising from tourism pollution and damage of tourism environment are often ignored, even by the administrative authority of tourism development. As such, it is very important and urgent to strengthen promotion of tourism environment protection and work out relevant policies and measures.

9.2 Factors Damaging Tourism Resources and Deteriorating Tourism Environments

The deterioration and damage of tourism resources and tourism environment are generally caused by many factors. Study on these factors can help us find countermeasures and play an active role in protecting tourism resources and environments in the development of tourism. The channels leading to damage of tourism resources and environments will be described from the following three aspects.

9.2.1 Damage of Tourism Environments Resulting from Undue Emphasis on Production

Pollution caused by industrial production to the natural eco-environment usually has a dreadful effect on tourist areas. For example, the Acapulco Beach in southern

Mexico used to be a resort with clear water and green hills as well as soft sands for vacation, summering and bathing. However, such a beautiful and charming beach has turned into a filthy and sloppy pit after it suffered history's worst pollution in recent years. Survey results show that daily wastes dumped into the bay through various channels include 300,000 m^3 of sewage, 440,000 tons of solid wastes, 170 tons of boron nitrate, 7 tons of phosphate salt, 5 tons of cleaner and 500 tons of industrial oil. The overspreading contaminants led to infection of many tourists by typhoid, taraxis, skin disease, etc. According to incomplete statistics, 20 % tourists are affected every year. Again, China's Fuchun River, which is known as "Lifeline of Tourism in Zhejiang", used to have green and rippling water which is so clear that you can see the river bed. In recent years, a number of enterprises like small fertilizer plants, pesticide plants and paper mills have been established along the river. They discharge large quantities of sewage and waste residue into the river, turning the river into a mess with rusty surface floating with bubbles, poisoning a large number of fish and prawn. Farmers cannot but fetch water from miles away as the water in the river can result in abdominal pain, dizziness, etc. According to statistics, Fuyang County alone has 258 paper mills, which discharge 11,560,000 tons of sewage into the river every year. Waste residues piled on the south bank of Fuchun River are 0.5 km long and about 3,000 m^3 in volume. Tourists can smell strong pesticide when they reach by boat the dock of the scenic resort Thousand Island Lake on the upper reaches of Fuchun River. It turns out that Lakeside Pesticide Plant discharges over 10 tons of sewage to the lake every day, polluting the lake and resulting in mass mortality of fishes (Bian and Tang 1985).

Due to backward industrial production technologies in China, the overall environment of key tourism cities has also seriously deteriorated as a result of pollution by "three wastes". Take Beijing for example, the monitoring result in 1982 shows that the average monthly dust fall content in the urban and suburban areas is 29.1 t/km^2, smoke days (visibility less than 4 km) rise from 60 days in the 1950s to the current 170 days, and the suspending particulate is 840 μg/m^3, 30 times higher than that in London. Most tourism-related water bodies in urban areas are eutrophicated, with the transparency averaging 0.5 m. Take Hangzhou as another example; statistics show that there are over 100 chimneys around the West Lake, with the annual mean dust fall content reaching 400 tons and lake-bottom silts averaging 1.3 m. The water in the lake is green yellow and turbid all the year round, with the transparency of only 0.5 m.

The damage to tourism attractions by backward farming and production modes and excessive lumbering, quarrying and water fetching is not only serious but also irreversible. In Jinan—"City of Springs"—for example, the groundwater level falls sharply due to long-term extensive exploitation of deep underground water in the past, resulting in dry-up of the springs. 20 years ago, all around the Jiuhua Basin in Mount Jiuhua, Anhui, are exuberant trees, but in the 1970s, only a small forest was left, with all the other trees felled up or reclaimed for farming. The "fair enneapetalous flower" which looked like green trees hung from Galaxy in the past now shows only bare mountain ridges. The seashore of Sanya in Hainan is probably the most promising seashore resort in China. However, natural landscapes here have been seriously damaged in recent years, with forests felled and burned and grotesque mountains and unique sceneries damaged. Incomplete statistics show that there are now more than 3000 people of 133 quarrying teams still making damages to tourist areas in Sanya. In South Hainan Ridge east of the resort Yalongwan alone, more than 1000 people try their best to blast the ridge, leaving only a trail of destruction.

Actually, irrational production modes and production layout not only cause damages to tourism background environment, but also incur direct harm to key tourism resources. For example, the Xishan Scenic Area in Echeng, Hubei, used to cover 4522 mu but now it measures only 2495 mu after encroachment of 26 factories and mines. Over 75,000 trees have been felled in just a few years. More than 20 places of interest have been destroyed (Xia 1985), including Fu Chai Diaoyutai, Taokan Reading Hall and Taogong Well. In Luoyang, many factories are established in the famous Tomb Area in Mangshan through explosion, destroying thousands of tombs. Cement plant, kiln dust and coal pit have been built in the famous Peking Man Site at Zhoukoudian, Beijing, since the 1970s. In particular, the cement plant is built beside the Dragon Bone Hill, and the quarry is only 200 m away from the unearthed area of "Peking Man", turning precious fossils buried in the ground into ashes (Zhou 1989).

9.2.2 Environmental Problems Resulting from Tourist Activities

In contrast with other industries that cause environmental pollution, tourism is referred to as "smokeless industry", which means that tourism can create profits just like other industries but is free from environmental pollution. But in fact, the development of tourism may also have severe impact on the environment if it is badly managed. Inadequate understanding and relevant countermeasures may also cause irreversible damages to the ecological environment of tourist areas, and finally damage the foundation of tourism.

The most obvious tourism pollutions include sanitary sewage, stool and rubbish. For example, 100,000 waste cans are left behind in tourist areas in Japan every year. Expenses for dealing waste cans in National Parks of Japan alone are more than 300 m yen (Liang 1987) in a single year. Three scenic areas of Huangshan in China, including Spring, Yuping Building and Beihai, are reported to have discharged 77,500 tons of sewages in 1984, generated a total of 3000 tons of stool in 1985 and piled 3400 tons of household garbage and 3000 tons of coal residue along mountain roads and alpine zones in less than three years from 1983 to September 1985 (Li 1985). The collection and disposal of refuse in mountainous tourist areas is difficult and bothers administrative authorities. As mountains are deep and forests are thick, people are unable to clean up and carry out these refuses but let them pile all over the mountains, which not only pollutes underground and surface water, and thus has an effect on animal and plant life, but also causes damage to landscapes, making tourists feel disgusted. The discharge of sanitary sewage and garbage, swimming and fuel-powered sightseeing boats result in the aggregation of toxic substances and eutrophication in the water. Water eutrophication means that nutrients such as nitrogen and phosphor needed by organisms enter the water in great quantities, and aquatic life reproduces, dies and decays in great number, causing reduction of dissolved oxygen in water, and deterioration and rancidity of water and generation of cancerogenic substances like nitrate, nitrite, etc. Eutrophication will form the cycle of nutrients, which is hard, costly and time-consuming to treat. The eutrophication area of Erie Lake located between the US and Canada is 5200 km^2, with the expected expenses for thorough treatment reaching US$40 billion. Almost all tourism-related water bodies in China have been polluted to different degrees. In particular, indexes like transparency, smell and dissolved oxygen of water bodies in quite a number of tourist areas exceed the standard, and floating objects, suspended materials and filth with oil stains make the visual impression extremely bad, which is very serious in Beijing, Suzhou, Hangzhou or even Lijiang River.

Extensive tourist activities may also directly undermine the balance of natural ecosystems in tourist areas. Lots of tourists stamp the soil flat, making the soil hardened and thus resulting in death of old trees, which is very common in famous parks in Beijing, Suzhou, etc. Tourists climb mountains, quarry, and damage the stable forest floors and humus layers developed over a long time under natural conditions, resulting in soil erosion, exposure of tree roots and drooping of grass, which all pose great threats to natural ecosystems in tourist areas. In addition, due to insufficient publicity and education, it is also quite common that tourists break branches, harm flowers and hunt animals in tourist areas. Although most tourists do not do things like this, the destruction caused by such conducts to crowded tourist areas is also terrible. Mountain tourist areas also have a very big hidden trouble, i.e. fires, most of which are caused by smoking and picnics of tourists.

The impact of extensive tourism development on the environment of key tourism cities is also the main cause of deterioration in tourism environment. Population of most tourism cities in China has been expanding in recent years. Population density has become very high and, with the rapid development of tourism and sharp increase of floating population, the already overstrained urban infrastructures fall far short of the surging need for energy, water, transport, building and municipal facilities. Meanwhile, the pollution load like sewage, waste gas, noise and rubbish also grows accordingly. Overcrowding and severe pollution inevitably result in worsening urban tourism environment.

9.2.3 Destructive Behaviours During Tourism Development and Urban Construction

Destructive behaviours refer to construction projects which are established in tourist areas for the purpose of exploitation and utilization, but which actually cause damages to tourism resources and tourism environment, especially tourism atmosphere environment. Such destruction is usually represented by facelift of historical sites and incongruity between new projects and sceneries in scenic areas, which as a result changes or even completely damages the original history, culture, national style and atmosphere in tourist areas that should be preserved. Almost all tourist areas have such a problem due to our lack of effective macro administration on tourism areas for some time. Take for example the famous Beijing Badaling Great Wall, where more and more hotels, shops and restaurants are built inside and outside the gate. These buildings stretch row upon row to the foot of the Great Wall, with the relative altitude higher than the Great Wall. Standing on the wall, tourists can only see a scene just like cityscape, rather than the most famous impregnable pass of the Great Wall. Such condition of Badaling Great Wall has raised attention and concern of experts, scholars, tourists and relevant departments.

Destruction to environment atmosphere in tourist areas is more prevalent in urban construction, as represented by Beijing. The careful design and general planning 500 years ago turned Beijing into an internationally rare monolithic architecture. It is a masterpiece in itself, vivid and symmetrical, and is rated as "the greatest individual project on the surface of the earth" in the history of architecture. From the perspective of tourism development, the style of this old city is in fact an attracting tourism resource. But now city walls and many historic buildings have given way to modern matchbox-type building clusters which are quite common in the world. Other examples are modern buildings around Hangzhou West Lake, high-rise buildings in Guilin, mansions and houses around Shenyang Imperial Palace.

9.3 Protection of Tourism Resources and Tourism Environments

According to the above analysis on factors leading to deterioration in tourism resources and tourism environment, the following countermeasures should be taken at present.

9.3.1 Strengthen Theoretical Research and Universal Education on Protection of Tourism Environments

Like other matters, the first issue to be resolved is understanding. Along with the continuous development of modern industry and sharp increase in global population, man brings greater interferences to the environment during the utilization of natural resources. The worsening natural environment, exhausting natural resources and recurrent severe revenges of nature for interference have finally made man understand that he, living in the natural environment, cannot do without nature. Continued worsening of the natural environment will eventually jeopardize the survival of humanity itself.

As such, people are now paying more and more attention to environmental science.

However, for a long time, environmental science research has been oriented to man's health needs but rarely addresses people's spiritual and psychological needs. This research is now illustrated in the subject of tourism environment protection. However, tourism is now still an emerging industry and research on protection of tourism environment has barely begun. To make tourism develop soundly, it is very important to draw a lesson from industrial development and strengthen theoretical research and social education on tourism environment. Current research contents include the following aspects:

1. Research on tourism resources, tourism environment and tourism relations includes the dependence of tourism on tourism environment, the constraining degree of environmental capacity to tourism development and destructive action of tourism development on tourism environment.

2. Methodological research on protection of tourism environment. The first is to determine the quality standard system of tourism environment, which should serve as the objective of the construction of tourist areas and standard for quality evaluation. Apart from general sanitation and environment quality standards, this system should also specify the quality standard of landscape aesthetics, quality standard of natural ecology and environmental quality standard meeting the needs of special tourist activities such as mountaineering, hunting and water sports. In addition to pragmatic and scientific principle requirements, this index system should also have qualitative, semi-quantitative or even quantitative evaluation indexes. The second is how to evaluate environments in tourist areas. Not only should the status quo of environments in tourist areas be evaluated, but also pre-evaluation on environmental impact of newly built tourist areas should be conducted. Meanwhile, problems such as principle and method for designing the environments in tourism areas should also be solved.

3. In the research on tourism environment protection works, protection of tourism environments is a complex system engineering involving many disciplines and specific technical problems. For example, we should build many new buildings without doing any harm to historical cities and styles of ancient building complexes; build necessary roads, bridges and even cableways in nature-based tourist areas while minimizing harms to local natural ecosystems and existing landscapes and solving problems such as discharge of sewage and pollutants in tourist areas; work out specific methods of restoring and maintaining heritages. Tourism is a cultural undertaking, which poses high-than-normal requirements for protection of tourism environments from the perspective of aesthetics and tourism psychology.

4. Research on policies for the protection of tourism environment works out relevant policies and rules according to general issues relating to tourism environments in a period, to provide decision-makers with consulting on countermeasures. As tourism is a multi-sector industry, its management is relatively complex and difficult compared with other industries, making researches on policies for tourism environment management all the more complex and important.

It is also necessary to greatly step up disseminating knowledge about tourism environment protection while conducting research and specific implementation on tourism environment protection. As tourist activities involve many travellers, conscious and concerted efforts of all travellers, tourism employees and local residents are very important. It is necessary to make people understand that all the historical, cultural and natural tourism resources are the common wealth of the entire mankind. Therefore, in protecting tourism environments, we are responsible for our descendants, not only for ourselves. Good protection of tourism environments in tourist areas and extension of lives of tourist resources are in line with the long-term interests of local residents and tourism operators in tourist areas.

It should be noted that most damages to tourism environments in tourist areas are caused by people's ignorance or even well-intentioned actions. This requires us to disseminate specific knowledge about protection of tourism resources and tourism environments through various channels, which is at least as important as our specific engineering measures.

9.3.2　Value Pre-evaluation of Environmental Impact During Planning and Construction of Tourist Areas

Environment destruction is usually an irreversible process, and mistakes in development and construction of tourist areas are usually irreparable. As such, it is necessary to, from the perspective of protection, take into full consideration all possible environmental impacts arising from construction and operation from the very beginning of development and planning of tourist areas, bring forward relevant countermeasures and work out relevant measures, so as to minimize destruction to tourism environments. If the above objective is estimated as unachievable even if all possible measures have been taken, it is necessary to limit the development and construction scale and number of tourists, or cancel some projects, or postpone development until conditions are favourable for solving possible environmental problems.

The contents of pre-evaluation of environmental impact include the evaluation of environmental capacity of tourist areas, evaluation of local economic capacity, evaluation of impact of development on population size of tourist areas and possible forecast of types of tourist source markets. Actually, in the planning and design of some large industrial and agricultural projects, especially in their feasibility analysis, the evaluation of environmental impact has become an indispensable content and should also be stressed in the development and planning of tourist areas, so as to reduce unnecessary losses.

9.3.3　Strengthen Assessment of Protection Scope and Environmental Capacity of Existing Tourist Areas

As the environment in tourist areas is especially important to tourism, every confirmed tourism area should have designated protection scope upon certification and assessment. Since all tourism objects occupy certain spaces, tourism development there entails limitation on other activities. The protection scope can be divided into several levels. The most basic scope is the area directly managed by tourist areas, or the minimum space required by main tourism objects and activities. Construction, restoration and utilization within that scope should be unconditionally in line with requirements for protection of tourism environments. The second is the protection scope established to protect harmony between the environment of tourist areas and surrounding landscapes and reduce the pollutions and damages caused by other activities to the background environment. All construction and production activities within that scope should be subject to some limitations. For example, the principle of maintaining a harmonious general environment and ensuring quality should be followed in connection with the types, scales, styles and capacities of buildings and possible discharge of "three wastes". Finally, protection in its broad sense should also be extended to the whole administrative area that rests upon tourism development. In the whole area, whether in industrial and agricultural projects, or construction of towns and roads, the protection of tourism environment is an important factor that should be considered from the time of planning.

It is also an important work to determine the best environmental capacity and tourism scale of the existing tourist areas. China has a huge population, so even though it is not highly developed in tourism, almost all tourist areas are overcrowded. Such overcrowding will not only directly undermine tourism environment, but also makes tourists unable to feel the due atmosphere and appeal of tourist areas, or even makes them feel uncomfortable and dissatisfied, which compromises the reputation of the tourist areas and will ultimately harm the future development of tourism. Switzerland, a country famous for highly developed tourism, has been aware of this since very early and put forward in 1980: "Keep cultural and historical traditions, pay attention to environment protection, and make development by focusing on improvement of service quality rather than quantity", i.e. the so-called Switzerland Tourism Concept (Li 1985). This concept is worthy of our reference.

The key to determining the best environmental capacity is determining the best tourist capacity, which is a complex question restricted by many factors, such as the natures and spaces of tourist areas, bearing capacity of natural environment, municipal engineering construction speed, regional economic development level, supply capacity of agricultural by-products, quantity and quality of labour, especially traffic conditions, such as the restriction on access of tourists to inner roads, restriction on access of tourists to dangerous inner passages, as well as other special requirements for resource maintenance. Beijing Forbidden City has been overloaded with tourists for years, so that the world-class famous museum is like a downtown area or a common recreational park. Many old trees are on the verge of death because their surrounding land is trampled solid by overcrowding tourists. Now, the lost "gold brick" in the grand hall, which can be regarded as a national treasure, has been ground dented. In particular, many tourists treat Forbidden City as a common park and often sit or lie there and even litter there, so all due environmental atmosphere there is completely lost. Many experts and scholars have long called for a termination of such a condition, but no measures have been taken for long time due to various restrictions. This problem has finally been resolved recently through adjusting the admission price of Forbidden City and restricting the number of tourists, which is a very good start.

The determination of environmental capacity in tourist areas, restriction on the reasonable scale of tourism development and proper treatment of the relationship between domestic tourism and international tourism are not only the needs of protection of tourism environment, but also the needs ensuring sound development of tourism in China. However, while taking relevant measures, it is also necessary to pay attention to the feelings of domestic tourists and appropriately address their needs for rest, recreation and travel.

9.3.4 Improve Legislation on Tourism Environment

Legislation refers to behavioural norms for tourists and tourism operators, obligatory interfere and punishment for destructive acts, as well as education and publicity for most people. The legislation contents should in particular include methods and rights for examination and approval of construction projects in tourist areas, provisions on protection scope and contents of tourist areas, and measures of punishment on people violating protection clauses.

In the early period after founding of new China, China promulgated *Measures for Protection of Historical Sites, Precious Cultural Relics, Books and Rare Creatures* in May 1950, issued *Administrative Measures for Protection of Local Cultural and Historical Places of Interests* in May 1951 and established the policy of actively protecting the cultural heritage and precious national treasures. Although these measures were not promulgated from the perspective of protection of tourism resources, they actually played such a role. Later, the State Council promulgated 18 clauses of Provisional Regulations on the Protection and Control of Cultural Relics in March 1961, and announced the first batch of 180 key cultural relics under state-level protection. In view of the serious damages to historical relics in "the cultural revolution" and years afterwards, the Standing Committee of the Fifth National People's Congress of the People's Republic of China also approved *Law of the People's Republic of China on the Protection of Cultural Relics* at its twenty-fifth session in 1982. According to the Law, "all institutions, organizations and individuals have the obligation to protect cultural relics of the state" and any disruptive behaviours "shall be investigated for criminal responsibility according to law". Later, China also promulgated various regulations like Forest Law and Law on the Protection of Rare Animals.

As people become more and more aware of environment protection, there are increasing requirements for managing three wastes and protecting the environment, especially the tourism development, and thus, legislation on protection of tourism environment has become urgent and imperative. On 7 June 1985, the Chinese Government enacted *Provisional Regulations on the Administration of National*

Parks and the Ministry of Urban and Rural Construction and Environmental Protection promulgated *Measures for the Implementation on Administration of National Parks*, the tourism-related Environmental Protection Law, etc.

Some other countries and regions have also laid down regulations on landscape and environment protection in specific tourist areas. For example, the USA enacted provisions on its first national park Yellowstone in 1872 (Sun 1988), and Japan enacted *Basic Law on Tourism* in 1963, which specifies: "protection, cultivation and development of tourism resources" should be one of the eight policies that must be implemented.

9.3.5   Improve Systems of National Parks and Reserves

In view of the serious damage to tourism resources and the need of protection of tourism environment, the State Council of China approved and forwarded the request of the Ministry of Urban and Rural Construction and Environmental Protection, Ministry of Culture and National Tourism Administration for instructions on examining and approving the first batch of national key scenic resorts in November 1982, and later announced the first batch of 44 national key scenic resorts. This is an important event in the protection of tourism resources and environment in China.

Key scenic resorts approved by the state actually have the nature of national parks and tourism environment reserves. Their establishment is similar but not identical to that of nature reserves. The concept of setting up nature reserves is mainly based on providing people with natural "background" of natural ecosystems by taking advantage of reserves, various ecosystems and natural store of biological species. Meanwhile, nature reserves also have the function of protecting natural vegetations and ecosystems formed by them and improving environmental condition. The establishment of reserves has placed a role in protecting tourism environments, especially natural scenery resources. Actually, along with the development of reserves, some scenic areas open to tourists are also included into the protection scope gradually. Nature reserves begin to diversify in type, or the same nature reserves have many functions. In China, the nature reserve fall roughly into the following types: for protecting typical and integral natural ecosystems, for protecting rare animal and plant resources, for protecting special landforms and typical geological profiles and for protecting natural sceneries. In particular, nature reserves aiming at protecting natural sceneries are the same as key scenic resorts or national parks. As such, some other countries have established natural conservation systems to give play to different functions of different reserves. For example, Romania divides natural conservation systems into national parks, natural parks, nature reserves, scientific reserves, scenic reserves, natural heritages, etc., and stipulates strict protection measures for each reserve. In China, it is imperative to adjust the relationship between nature reserves and scenic resorts and to establish natural conservation systems.

The approval and announcement of key scenic resorts merely provide a good basis for the protection work of tourism environments in these reserves. Moreover, according to requirements of the state on the management of scenic areas, it is also necessary to carefully plan scenic resorts, improve organizations, strengthen leadership and implement effective unified management system, and actively and steadily push forward the development and construction work of scenic areas while strictly protecting the landforms, vegetation, natural ecology and historical and cultural relics within scenic areas. Key construction projects must be subject to examination and approval by upper competent authorities. Scenic areas should be opened to tourists in a planned manner according to their specific conditions, in order to develop tourism scientifically and reasonably.

The continuous announcement of national parks (i.e. national key scenic resorts) and local parks, as well as the constant improvement in nature reserve systems of the state in future, especially the continuous improvement in actual works of various reserves, is sure to create a new look on China's tourism environment and protection of tourism resources and therefore provide a solid foundation for the sound development of tourism in China.

References

Bian Q., & Tang, Z. (1985, Aug 13). *How can Fuchun River stand to so much pollution*. China Tourism News.
Li, Z. (1985, Sept 24). *Maintain tradition and assure success by quality*. China Tourism News.
Liang, S. (1987). *Main "tourism hazard" in Huangshan scenic area and countermeasure research, collected essays on tourism* (4th ed.). Beijing: Tourism Times Agency.
Sun, X. (1988). *Investigation report of U.S. national parks, research on tourist attraction* (p. 565). Shanghai: Tongji University Press.
Xia, P. (1985, June 25). *It is urgent to rescue Xishan scenic resorts*. China Tourism News.
Zhou, Y. (1989). *Environmental issue of Beijing tourism and countermeasures, Beijing tourism development strategy* (p. 209). Beijing: Beijing Yanshan Press.

Chapter 10
Geoparks

As an innovative type of tourism destination in China emerging at the very beginning of the twenty-first century, geoparks are landmarks in the history of tourism earth-science (Chen 1998) functioning as newborn children in a new era. The birth of geoparks created a solid foundation for tourism earth-science in both theoretical research and policy practice, elevating tourism earth-science to a new level and demonstrating its shining and sustainable future. Reflecting the significance of geoparks in knowledge building for tourism earth-science, and as suitable places for visitors to learn earth-sciences, a chapter on geoparks has been included in this new edition of *An Introduction of Tourism Earth-science*. How geoparks came into being in China, the purposes of geopark establishment, criteria for nomination, approval procedures and criteria for final designation will be dealt with in this chapter, with a brief introduction about current developments and future prospects of UNESCO Global Geoparks.

10.1 Historical Evolution of China's National Geoparks

10.1.1 The Birth of Geoparks

The Chinese Academy of Tourism Earth-science and Geopark Research was established in Beijing in April 1985, when Chen Anze was elected as its president. At the inaugural meeting, several policy recommendations were proposed to the State Council of China, emphasizing the importance of earth-sciences surveys and research in the development of the tourism industry. These policies state that earth-science parks and tourist museums promoting earth-sciences should be developed in natural beauty-based tourism destinations such as Mount Lushan, Mount Huangshan, Mount Wuyi, Mount Taishan, Three Gorges, Guilin, Wudalianchi and

© Springer-Verlag Berlin Heidelberg and Science Press Ltd. 2015
A. Chen et al., *The Principles of Geotourism*, Springer Geography,
DOI 10.1007/978-3-662-46697-1_10

Zhangjiajie, so that the scientific values of tourism destinations, or science-oriented tourism, could be promoted.[1] In this manner, tourism earth scientists created the new term and concept of an earth-science park. With an associate president from the Chinese Society of Tourism earth-science chairing the November 1985 Zoning Conference on Geo-based Natural Heritage Reserves at Zhangjiajie National Forest Park, Hunan (sponsored by the Ministry of Geo-mining Industry), the proposal from the first meeting of the society was revisited via the planned creation of earth-science parks. Further, a site visit was undertaken to Zhangjiajie National Forest Park. Although designated as a forest park, the main attraction in the core zone of Wulingyuan is landforms consisting of rocky columns shaped from Devonian quartzitic sandstone, rather than forest scenery. Accordingly, delegates of the conference proposed to the Ministry of Geo-mining Industry that Wulingyuan National Geopark should be created,[2] and hence, national geoparks were born. The Ministry of Geo-mining Industry was so interested in the idea that it issued a notification to promote the establishment of tentative geosite nature reserves, of which geoparks were considered one type.

Geoparks were classified as a form of geosite reserves in the 1995 *Regulations on the Protection of Geo-sites* issued by the Ministry of Geo-mining Industry.[3] However, for various reasons, none of the 86 geosite nature reserves (12 national reserves, 33 provincial reserves, 9 prefecture-level reserves and 32 county-level reserves) were designated as geoparks by the end of 1999. Against this background, President Chen Anze of the Chinese Society of Tourism Earth-science repeatedly advised the head official of the Ministry of Geo-mining Industry in charge of geosite administration, in anticipation of the early creation of geoparks, although the positive answer was never acquired. A geosite reserve differs from a geopark. The major purpose of a reserve was to protect geosites, which is a core responsibility of the Ministry. In contrast, a park as a public social service provider could conserve geosites while hosting tourists, emphasizing sustainable use of geosites for science promotion and tourism services. Although there was a recommendation to create earth-science parks, geoparks or national geoparks as early as 1985 by tourism earth scientists, the dream was not transformed into reality despite Chinese government efforts in subsequent years (Chen 1996).

In 1998, the Ministry of Geo-mining Industry became the Ministry of Land and Resources (MLR). Geosite administration became an important part of geoenvironment protection, so this ministerial reorganization afforded the opportunity for the Chinese government to create geoparks. At the same time, the UNESCO earth-science Sector began to promote the creation of geopark networks and enhance

---

[1] The 25th anniversary memorabilia of tourism earth-science. Tourism Earth-science and Geopark Studies Research Branch in Chinese Society of Geological Science, 2010-9-18.

[2] Corpus of Geological Nature Reserve Regionalization and Wulingyuan Scientific Exploration, 1985.

[3] *Regulations on Geosite Conservation Management*. Ministry of Geo-mining Industry Order #21, 1995-5-4.

geosite conservation. Chen Anze once again proposed a geopark network in China, which was promptly accepted by delegates at the National Conference on Geolandforms and Landscape Protection held by the MLR in December 1999. In April 2000, the Minister finally officially accepted the idea of creating geoparks.[4] On 24 August 2000, notification of the organization and formation of the Leadership Group of National Geo-heritage (Geopark) was issued by the MLR.[5] The group's members include representatives from the Ministry of Finance, National Administration of Environmental Protection, China Geological Survey Bureau, Ministry of Construction, Chinese Academy of Geological Sciences and Chinese Society of Geological Sciences. On 22 September 2000, the MLR issued its *Notification to Nominate National Geoparks*[6] and outlined regulations regarding nomination rules, evaluation criteria and approval procedures, opening the way for the central government to create Chinese national geoparks.

In the year 2000, nominations for geoparks were made by 13 local governments, and eventually 11 geoparks, including Shilin National Geopark in Yunnan, were designated following a rigorous selection process by the National Geopark Evaluation Committee. These represent the first national geoparks in the world, creating a new era of geoparks and marking 2000 as the Creation Year of Geoparks in China.

10.1.2 The Definition of a National Geopark

Just as its name implies, a geopark is a natural park with major attractions being natural land and water formations. According to the *Notification to Nominate National Geoparks*, a geopark is a science-based park with special scientific values, rare natural attributes, superb aesthetic values and a thematic geolandscape of a proper size and distribution range; with ecological, historical and cultural values integrating its natural and cultural landscapes; with a purpose to conserve geosites and support sustainable development with respect to the local economy, culture and environment; and providing a public service by promoting highly scientific appreciation via sightseeing tours, holiday recreation, health promotion and recovery, science promotion and education, and other cultural leisure activities. At the same time, a geopark is an important protected area of an ecologically sound environment, and a home base for geoscience research and promotion.

---

[4] See *Notification of Request from Minister of National Land and Resources* #29, Direction on the request geosites conservation. 2000-3-3.

[5] See *Notification on the Formation of the Leadership Group to Designate National Geosites* (*Geoparks*), Ministry of Land and Resources Official Document No. [2000] 68.

[6] A *Notification on the Nomination of National Geoparks*, Ministry of Land and Resources Official Document No. [2000] 77.

10.1.3 The Mission and Purpose of China's National Geoparks

The Chinese government's mission with respect to creation of geoparks can be summarized as follows:

1. The first mission is to conserve geosites and protect the natural environment. Geosites were formed by internal and external geological agents, creating landforms in the earth's surface with aesthetic values, and life records or tectonic records preserved in the rock layers with significant science values. Geosites provide evidence about the evolutionary history of the earth, with important and irreplaceable values of science; and accordingly they deserve rigorous protection. The establishment of geoparks is the ideal way to preserve geoheritage. Geoparks are special places in which mega-site engineering projects, mining activities and urbanisation are prohibited in order that the natural environments within can be effectively protected. The main reason to establish geoparks is conservation and protection.

2. The second mission is to disseminate knowledge of earth-sciences and increase public understanding of sciences more generally. Geoparks are filled with typical geosites that represent an extremely rich information source regarding earth's evolutionary history and as such these parks can function as natural classrooms to disseminate knowledge of earth-sciences to tourists. For this purpose, each geopark is required to create facilities for science interpretation and provide a number of interpretive staff so that tourists will gain earth-science knowledge and a better understanding of sciences during a leisurely tour of the site.

3. The third mission is to activate geotourism and promote sustainable regional development of the local economy and society. As an important branch of economic development, tourism has become the largest industry in terms of production. Opening to the public through geotourism is one of the major purposes of geoparks, which are substantially different from the previously established geosite reserves. Tourism is deemed a good way of boosting employment for local communities in the vicinity of the park and facilitating local economic development. Most geoparks are located in mountainous regions far from city centres, with rare resources and under-developed economies. The establishment of a geopark can facilitate economic development for impoverished locals in such regions. Notable economic development in most regions where geoparks are located vividly testifies to the importance of this particular purpose in China.

Geoparks are serving the purpose of protection of geosites and natural environments, and at the same time, are utilizing geosites for disseminating scientific knowledge to the public through tourism activities and promoting economic development. In this sense, the geopark idea is a comprehensive way of providing benefits to the public for future generations.

10.2 China's National Geoparks

10.2.1 The Emblem of China's National Geopark

The emblem of China's national geopark is shown here. The idea of this emblem was proposed by Chen Anze, and its design was undertaken by Liang Xiangrong. The upper part of the outer rim is inscribed with Chinese characters for the National Geopark of China, with their English translation in the lower part of the rim. The inner circle is a pictographic symbol in which the upper part symbolises the ancient Chinese character for Mountain, representing natural scenery containing unique peaks, strange caves and rocky stone formations; the middle part symbolises the ancient Chinese character for water, representing waterscapes and accumulated strata or geological folds and faults from tectonic movements; and the lower part depicts the dinosaur *Mamenchisaurs* discovered in the Jurassic stratum in Sichuan. Displaying various elements of geological landforms and landscapes, and symbolizing the rich tradition of the Chinese culture, the emblem combines science and culture, using meaningful, concise and living images to demonstrate the unique Chinese civilization.

10.2.2 Categorization of China's National Geoparks

Geoparks in China are a series of parks with different administrative tiers and various geo-site contents, varying in size and geographical diversity. These parks can be classified in a number of different ways as follows:

1. Classification by the administrative level at which the park was designated:
   (a) UNESCO Global Geopark: approved and awarded with a certificate by UNESCO.
   (b) National Geopark: approved and awarded with a certificate by the central government (represented by the MLR).

(c) Provincial Geopark: approved and awarded with a certificate by the provincial government (represented by the Department of Land and Resources).

(d) County Geopark: approved and awarded with a certificate by the county-level government.

2. Classification by planned area size as follows:
   (a) mega: >500 km^2
   (b) large: 101–500 km^2
   (c) medium: 21–100 km^2
   (d) small: <20 km^2.

3. Classification by major science-based tourism function as follows:
   (a) Geoparks for scientific research and exploration contain geosites with substantial value for scientific research, and possibly modest aesthetic value. The major mission of these parks is to preserve rare geosites; and core zones are open only to professional researchers. Visitors in other zones within the park are typically students in the earth-sciences.
   (b) Geoparks for appreciating natural beauty and joining sightseeing tours have a certain degree of scientific value, but the geolandscapes have more notably high value for aesthetic appreciation, and hence, the essence of such geoparks is its strong attraction for general visitors.

4. Classification by geolandforms and geolandscapes. The geological landform and landscape (or geolandscape) is a natural resource of extreme significance and a tourism resource of great importance, as well as the major geopark attraction. In terms of geological processes and landscape categories, geoparks can be classified into seven major classes, 25 medium classes and 56 minor classes (Table 10.1). The classification of geoparks with respect to geological

Table 10.1 Classification of geoparks and geo-sites

Major class	Medium class	Minor class
1. Geological sections	1. Stratum	1. Global profile (Golden Spike)
	[1]	2. National profile
	[1]	3. Regional profile
	[1]	4. Local profile
	2. Magmatic	5. Basic and ultrabasic rock
	[1]	6. Neutral rock
	[1]	7. Acidic rock
	[1]	8. Alkaline rock
	3. Metamorphic	9. Contact aureole
	[1]	10. Thermodynamic metamorphism
	[1]	11. Migmatization
	[1]	12. High or ultra-high pressure metamorphism
	4. Sedimentary rock	13. Sedimentary rock sections

(continued)

Table 10.1 (continued)

Major class	Medium class	Minor class
2. Tectonics	5. Tectonic evidence	14. Global tectonics
	[1]	15. Regional tectonics
	[1]	16. Local tectonics
3. Palaeontological	6. Palaeoanthropological	17. Palaeoanthropological fossils
	[1]	18. Palaeoanthropological activities
	7. Palaeozoological	19. Palaeoinvertebrates
	[1]	20. Palaeovertebrates
	8. Palaeobotanical	21. Palaeobotanical
	9. Palaeontological	22. Palaeontological activities
4. Mineral deposits	10. Typical minerals	23. Typical minerals
	11. Typical ore deposits	24. Metal ore deposit
	[1]	25. Non-metallic deposit
	[1]	26. Energy source deposit
5. Landforms	12. Rocky	27. Granite
	[1]	28. Clastic rock
	[1]	29. Karst
	[1]	30. Loess landform
	[1]	31. Sand-accumulated landform
	13. Volcanic	32. Volcanic edifice
	[1]	33. Lava
	[1]	34. Pyroclastic debris
	14. Glacier	35. Glacier exaration
	[1]	36. Glacier accumulation
	[1]	37. Periglacial
	15. Fluvial	38. Fluvial erosion
	[1]	39. Fluvial accumulation
	16. Marine changed	40. Marine abrasion
	[1]	41. Marine accumulation
	17. Tectonic	42. Tectonic landform
6. Waterscapes	18. Springs	43. Hot springs
	[1]	44. Cold springs
	19. Limnological	45. Lakes
	[1]	46. Wetlands
	20. Rivers	47. Riverways
	21. Waterfalls	48. Waterfalls

(continued)

Table 10.1 (continued)

Major class	Medium class	Minor class
7. Evidence of geological events	22. Earthquake sites	49. Palaeoearthquake sites
	[1]	50. Modern earthquake sites
	23. Meteorite impact	51. Meteorite impact
	24. Geological disasters	52. Hill avalanche
	[1]	53. Landslide
	[1]	54. Debris flow
	[1]	55. Land subsidence
	25. Mining sites	56. Mining sites

causes is essential to understand their characteristics and attributes. However, a diverse range of geolandscapes can often be found in a large geopark because geological phenomena co-occur within a particular region. Accordingly, the practice of classification or naming of a park should be conducted in terms of its major geolandscapes.

10.3 The Nomination of China's National Geoparks

10.3.1 Criteria to Nominate a Geopark

The geosites in a proposed national geopark must display national representativeness and have nationally or internationally unique values with respect to science, education and aesthetic appreciation. These criteria are as follows:

1. Uniqueness of geosite resources: geosites that provide important geological evidence to demonstrate a major geological event or an evolutionary stage of the earth's history in a mega-region or in a global sense; geolandscapes or phenomena of international or national significance having typical geoprofiles and fossils in the context of international or national mega-region strata; rare examples of geosites in the national or international context.
2. Quantitative criteria, area size and value for science promotion: at least three geosites must possess the uniqueness of national significance, and a minimum of 20 geosites is required to serve science promotional and educational activities.
3. Values of aesthetic appreciation of the geosites: its attractiveness for general visitors will boost local tourism after geopark building, and facilitate sustainable development of the local society and economy.
4. Geosites have been effectively protected: ongoing or planned large-scale projects by the transportation, irrigation and mining industries that are relevant to local social and economic development must not threaten or destroy the geosites.
5. Two years later after the opening of a provincial geopark.

10.3.2 *Nominator of a Proposed National Geopark in China*

The country-level government where the provincial geopark is located submits an application to the Department of Land and Resources of the relevant provincial government. Through reviews of experts organized by the Department of Land and Resources, the qualified park continues to submit the application materials to the MLR.

10.3.3 *Nominating Date and Limits of Proposed Parks*

In principle, nominations would be accepted once every two years, with a maximum of two parks nominated in any individual province and only one park nominated in a province where more than 10 national geoparks have already been established.

10.3.4 *Nominating Material and Documentation*

The nomination materials and documentation the applicants must provide are as follows:

1. Application Form for National Geopark Nomination. The production of this form was overseen by the MLR. It includes the name of the geopark, its location, area size, current management and recommended improvements, major geosites and their conservation issues, natural environments and cultural landscapes, social and economic conditions of the park and its situated communities, scientific research activities, introduction of previous achievements and the master plan, placement of infrastructure, reviews from the expert panel, and comments from the applicant (Department of Land and Resources of the relevant provincial government).
2. An integrated survey report of the proposed national geopark. This has three major components: a brief description of the geopark, an evaluation of its geological context and geosites, and a report on current conservation issues in the geopark. The main purpose was to confirm the existence of and evaluate the geosites, other natural scenic resources and cultural resources situated in the park, to clarify why the park would be designated as a national geopark.
3. A master plan of the proposed geopark. The main purpose of the master plan was to outline how an area containing geosites and qualifying as a national geopark will be incorporated into a national park for public visitation. The major tasks include ascertaining the character of the park; delineating the park and its buffer zones for protection; confirming protection measures to conserve

the geosites and other resources situated in the park; documenting its overall spatial layout and setting up various functional zones; confirming its tourism carrying capacity and administrative and management measures to organize geotourism and sightseeing activities; making overall arrangements for public services, hospitality and other infrastructure; arranging the sequential order of development and construction, and calculating investment and income; and clarifying the park management mechanisms and supportive measures. The master plan is the foundation for national geopark creation, which will be reviewed and approved by the Department of Land and Resources, and revised within a certain period.

4. Maps, drawings and photographs. These include maps of park location, geological maps, topographical maps, satellite photographs, aerial photographs, maps of environmental geology and planning maps.
5. Videos, video compact discs and photo albums displaying the geosites and components to be conserved within the proposed geopark.
6. Official certification documents for approval of park establishment, including documentation approving the provincial geopark and a certificate of land use by the local authorities (in principle, the proposed national geopark must be a provincial geopark in which park building, management and operation has occurred for over two years).

10.4 The Evaluation and Approval of a National Geopark

10.4.1 Tentative Evaluation Criteria for a National Geopark (2000–2014)

The review of the criteria includes three components: natural attributes of the geosites, conservation feasibility and the management mechanism. These are specified into 12 items that are each awarded points according to their importance. Only those scoring at least 60 points qualify for designation:

1. Natural attributes (60 points) and values of the geosites, including their unique representation (15 points), scarcity (17 points), natural conditions (8 points), systematicness and integrity (10 points) and aesthetic values.
2. Conservation feasibility (20 points): the level of protection of the geosites and their social and ecological value, including feasibility of park area size (6 points), economic and social values (6 points) and ecological values (8 points).
3. Management status (20 points) with respect to the park management agency and its staff placement, including staff positioned (4 points), park boundary and land use authorities (3 points), surveys for basic information gathering (6 points), infrastructure and transportation (7 points).

10.4.2 The Proposed New Evaluation Criteria for a National Geopark

The tentative evaluation criteria are being revised since they were not designed to serve the current demand. The updated version has three components: geosite values, park achievements and park prospects. These are specified into 10 items and points will be given according to the importance of each. Only those marked with 60 points and more are qualified to be designated:

1. Geosite values (50 points), including scientific value (20 points); aesthetic value (20 points); and values relating to science promotion and education (10 points).
2. Park achievements (25 points), including integrity of park area, size and feasibility (5 points); geosite conservation status (7 points); scientific interpretation system building status (7 points); and infrastructure building status (6 points).
3. Park prospects (25 points), including tourism development conditions and prospects (8 points); role in social–economic development (7 points); and safeguarding measures from local government (mainly in management agency building and construction funding) (10 points).

10.4.3 Approval Procedures for National Geoparks

A national geopark designation is extremely valuable because of the rigorous approval process for proposed parks, as follows:

1. Review by the expert committee for national geoparks. The reviewing committee consists of 30 experts appointed by the MLR. After reading the application materials, there is open expert voting on how points should be awarded. Only those applications scoring at least 60 points and approved by more than 2/3 of the voters are qualified to be designated by the National Geopark Leadership Group in the MLR.
2. The approval of the National Geopark Leadership Group in the MLR. Following evaluation by the reviewing expert committee, the list of qualified parks would be submitted to the group and the list of the final approved national geoparks would then be publicized.
3. National geopark building time limits and the final approval for formal designation. According to the park building criteria, a proposed national geopark must be developed within three years of qualifying to undertake geopark operation. Once the initial phase of park building has been completed, the local government would apply for final approval via the Department of Land and Resources of the relevant province. The National Geopark Leadership Group would dispatch experts to decide whether the park qualifies for approval following initial review by experts from the Department of Land and Resources

in the provincial government. A report would be prepared by experts from the MLR and submitted for final approval by the Ministry. Based on the recommendation of the expert group, the Ministry would make the final decision on whether the formal title of National Geopark is conferred. The Ministry would issue a formal notification to confirm the final approval, and the approved park would choose the appropriate time to begin formally using the title of National Geopark and open the park to the general public. A proposed geopark would lose its qualification to become a formal national geopark if the park building were not completed within the three planned years.

10.5 Working Criteria for Building National Geoparks

Not only are a number of national level geosites required for a national geopark, various park building efforts are necessary for the park to host tourists as well. Park building contents can be specified into 14 items, which each score points according to their importance, as follows.

10.5.1 Geopark Planning and GeoSite Conservation (30 Points)

1. Geopark planning (10 points) is the foundation on which a geopark is built and managed. Any geopark is required to undergo park planning according to the *Technical Rules for Developing a National Geopark Plan*.[7] The completed plan would be reviewed by an initial expert group organized by the Department of Land and Resources in the provincial government, and final approval is given by the MLR before it is publicized by the local government for implementation.
2. Geosite conservation (10 points): geosites are the essential feature around which a geopark is established and also represents the major conservation component. Every park must develop an inventory of its geosites and relevant archival records after a thorough survey of type, distribution, number and grade of all the geosites within the park. Relevant, detailed and effective measures should be developed and implemented, under the responsibility of the particular sector and individuals.

---

[7] *A Notification of Technical Rules for Developing a National Geopark Plan*, Ministry of Land and Resources Official Document No. [2010] 89.

3. Geopark demarcation (10 points). The boundaries of a geopark must be surveyed and demarcated on the basis of coordinates at inflexions, and border marker posts erected. Land use authorities must be clear that any mining and commercial prospecting is prohibited.

10.5.2 Scientific Interpretation System Building in Geoparks (40 Points)

As a major component of a geopark, creating a scientific interpretation system is of major importance for earth-science knowledge dissemination to tourists, and in the development of environmental education. Scoring for this attribute refers to:

1. The main and ancillary monuments (5 points). The main monument must be erected at the most representative section of the park and must contain the name of the park, approving organization, date of approval, National Geopark of China emblem, a map of the park and its concise description. In principle, an ancillary monument may be erected at each individual section of the park, providing a map of the relevant section of the park and its concise description.
2. Geopark museum (10 points). A geomuseum shall be set up to explain the essential geosite features of the park, to disseminate geoscientific knowledge. The display area must be larger than 800 m^2 in principle, and a certain number of interpreters must staff the park.
3. Science promotion studio (5 points). A science promotion studio must be set up and employ modern, multimedia technology to showcase the evolutionary history of the geolandscapes in the park. This can be built as part of the museum or as an individual building, ensuring sufficient seating (for more than 80 people, in principle) to meet tourist demand.
4. On-site interpretive posts (10 points). Scientific interpretive posts must be erected near representative or unique geosites to help tourists understand the geoscientific significance of the site. The information must be presented with scientific precision and employ diagrams to ensure it is easily understandable. In principle, at least 50 such posts should be set up in a park, with a minimum of 30 in any individual park section.
5. Road signs (5 points). No fewer than three road signs must be erected along the relevant main roads so that tourists can find the park. The park's location and emblem must appear on each sign in a standardized manner and style.
6. Science tour guiding map and travel route for scientific exploration and science promotion (5 points). A concise, clear, direct and publication feasible science touring map must be developed according to the *Technical Rules of Developing a National Geopark Plan*, with the intention of formal publication in the future. Travel routes designed for scientific exploration and science promotion must incorporate typical geolandscapes and human landscapes within the park.

10.5.3 Research and Science Promotion Activities in Geoparks (15 Points)

1. Scientific research (7 points). A scientific research plan and short-term (3–5 years) action plan must be developed by the park agency. Research funding must be arranged and at least one research project implemented within the first year.
2. Science promotion activities (8 points). The park agency must develop a science promotion plan and short-term (3–5 years) action plan directed toward school children, college students and general visitors, and geological samples and park souvenirs. Further, science promotion reading materials (books and DVDs), such as geopark book series, must be developed and published.

10.5.4 Administration and Digitalisation of Geoparks (15 Points)

1. Management agency (5 points). A well-organized national geopark management agency (management bureau or management section) must be set up with the formal approval of the institution planning office from the county or higher-level government. The major leaders must be appointed formally, with functional departments, management offices and a management mechanism with scientific rationale.
2. Staff placement (5 points). Professional management staff must be employed, including some with background knowledge of tourism earth-science (3–5 persons) and a certain number of professional tour guides.
3. Geopark digitalisation (5 points). A geopark database and monitoring system must be established, and an independent website created and regularly updated.

10.6 Developmental Stages of Chinese National Geopark

China began to establish national geoparks from 2000 (Chen et al. 2007) and had approved 241 national geoparks in seven batches by January 2014. After 2009, approval methods were reformed, which applied from the fifth batch of nominations. Approval to construct a national geopark will first be awarded to applicants that have achieved the basic conditions of a national geopark, although the construction of a national geopark is required within three years. Geoparks can be proposed as formal national geoparks only after inspection and acceptance. Up to May 2014, 174 national geoparks had completed construction and been opened to the public; the remaining 77 are still under construction.

The seven batches of geoparks can be summarized as follows. The first batch, approved in October 2000, consisted of 11 national geoparks (including Shilin National Geopark in Yunnan) that were announced to the public in March 2001. The second batch was nominated in October 2001 and consisted of 33 national geoparks including Mount Huangshan National Geopark in Anhui, announced to the public in February 2002. The third batch of 41 national geoparks was proposed in November 2003 and included Mount Wangwu National Geopark in Henan, which were announced to the public in March 2004. The fourth batch was put forward in August 2005 and consisted of 53 national geoparks (including Mount Taishan National Geopark in Shandong) that were announced to the public in September 2005. The fifth batch, approved in August 2009, encompassed 44 geoparks including Mount Changbai National Geopark in Jilin, announced on 19 August, 2009 as having construction qualifications for national geoparks: Hongkong National Geopark was specially approved in September 2010. The sixth batch, which was approved on 13 November 2011, consisted of 17 geoparks include Luoping Biota Geopark in Yunnan announced on 30 December 2011 as having construction qualifications; and 19 geoparks including Zhangye Danxia Geopark in Gansu were declared on 23 April 2012. The seventh batch, which was approved on 21 December 2013, consisted of 22 geoparks including Enshi Tenglongdong Grand Canyon Geopark in Hubei, whose construction qualifications were announced on 9 January 2014.

To deal with the quantity of Chinese national geopark applications and to access information about the abovementioned geoparks, for this study, they were numbered according to the approval time and announced list by the MLR (see Table 10.2). For instance, 001 is the number given to Shilin National Geopark in Yunnan. These numbers are their permanent codes. Table 10.2 gives the number and name of every national geopark in every state.

Table 10.2 List of National Geoparks of China

Designated year	Number	Geopark name
2000	1	Shilin National Geopark, Yunnan
	2	Zhangjiajie Sandstone Peak Forest National Geopark, Hunan
	3	Mount Songshan National Geopark, Henan
	4	Mount Lushan National Geopark, Jiangxi
	5	Chengjiang Faunal Paleobios National Geopark, Yunnan
	6	Wudalianchi National Geopark, Heilongjiang
	7	Zigong Dinosaur National Geopark, Sichuan
	8	Zhangzhou Littoral Volcano National Geopark, Fujian
	9	Mount Cuihuashan Landslides National Geopark, Shaanxi
	10	Mount Longmenshan Tectonic Geology National Geopark, Sichuan
	11	Mount Longhushan National Geopark, Jiangxi

(continued)

Table 10.2 (continued)

Designated year	Number	Geopark name
2002	12	Mount Huangshan National Geopark, Anhui
	13	Dunhuang Yardong National Geopark, Gansu
	14	Hexigten National Geopark, Chifeng, Inner Mongolia
	15	Tengchong Volcano National Geopark, Yunnan
	16	Mount Danxiashan National Geopark, Guangdong
	17	Hailuogou National Geopark, Sichuan
	18	Shanwang National Geopark, Shandong
	19	Jixian National Geopark, Tianjin
	20	Dadu River Canyon National Geopark, Sichuan
	21	Dajinhu National Geopark, Fujian
	22	Jiaozuo Mount Yuntaishan National Geopark, Henan
	23	Liujiaxia Dinosaur National Geopark, Gansu
	24	Jiayin Dinosaur National Geopark, Heilongjiang
	25	Shihua Cave National Geopark, Beijing
	26	Changshan National Geopark, Zhejiang
	27	Laiyuan Mount Baishishan National Geopark, Hebei
	28	Mount Qiyunshan National Geopark, Anhui
	29	Qinhuangdao Liujiang National Geopark, Hebei
	30	Huanghe Hukou Waterfall National Geopark, (Shanxi, Shaanxi)
	31	Zhanjiang Huguangyan National Geopark, Guangdong
	32	Fuping Natural Bridge National Geopark, Hebei
	33	Anxian Bioherm-Karst National Geopark, Sichuan
	34	Zaozhuang Mount Xiong'ershan National Geopark, Shandong
	35	Zongyang Mount Fushan National Geopark, Anhui
	36	Yanqing Silicified Wood National Geopark, Beijing
	37	Neixiang Baotianman National Geopark, Henan
	38	Linhai National Geopark, Zhejiang
	39	Luochuan Loess National Geopark, Shaanxi
	40	Yigong National Geopark, Tibet
	41	Huainan Mount Bagongshan National Geopark, Anhui
	42	Chenzhou Mount Feitianshan National Geopark, Hunan
	43	Mount Liangshan National Geopark, Hunan
	44	Ziyuan National Geopark, Guangxi

(continued)

Table 10.2 (continued)

Designated year	Number	Geopark name
2004	45	Mount Wangwushan National Geopark, Henan
	46	Jiuzhaigou National Geopark, Sichuan
	47	Mount Yandangshan National Geopark, Zhejiang
	48	Huanglong National Geopark, Sichuan
	49	Chaoyang Fossil National Geopark, Liaoning
	50	Baise Leye Dashiwei Sinkholes National Geopark, Guangxi
	51	Xixia Mount Funiushan National Geopark, Henan
	52	Guanling Fossils National Geopark, Guizhou
	53	Beihai Weizhou Island Volcano National Geopark, Guangxi
	54	Mount Chayashan National Geopark, Henan
	55	Xinchang Silicified Wood National Geopark, Zhejiang
	56	Lufeng Dinosaur National Geopark, Yunnan
	57	BuErJin Kanas Lake National Geopark, Xinjiang
	58	Jinjiang Shenhuwan National Geopark, Fujian
	59	Yulong Liming-Mount Laojunshan National Geopark, Yunnan
	60	Qimen Guniujiang National Geopark, Anhui
	61	Jingtai Yellow River and Stone Forest National Geopark, Gansu
	62	Shidu National Geopark, Beijing
	63	Xingyi National Geopark, Guizhou
	64	Xingwen Shihai National Geopark, Sichuan
	65	Wulong Karst National Geopark, Chongqing
	66	Mount Arxan Volcano-Warm Spring National Geopark, Inner Mongolia
	67	Fuding Mount Taimushan National Geopark, Fujian
	68	Jianzha Kanbula National Geopark, Qinghai
	69	Zanhuang Zhangshiyan National Geopark, Hebei
	70	Laishui Yesanpo National Geopark, Hebei
	71	Pingliang Mount Kongtongshan National Geopark, Gansu
	72	Qitai Silicified Wood-Dinosaur National Geopark, Xinjiang
	73	The Three Gorges of Yangtze River National Geopark, (Hubei, Chongqing)
	74	Haikou Mount Shishan Volcano Cluster National Geopark, Hainan
	75	Mount Xishan National Geopark in Taihu Lake, Jiangsu
	76	Xiji Huoshizhai National Geopark, Ningxia

(continued)

Table 10.2 (continued)

Designated year	Number	Geopark name
	77	Jingyu Volcano-Mineral Water National Geopark, Jilin
	78	Ninghua Swan Karsts National Geopark, Fujian
	79	Dongying Yellow River Delta National Geopark, Shandong
	80	Zhijindong Cave National Geopark, Guizhou
	81	Foshan Mount Xiqiaoshan National Geopark, Guangdong
	82	Suiyang Shuanghedong Cave National Geopark, Guizhou
	83	Yichun Granite Stone Forest National Geopark, Heilongjiang
	84	Qianjiang Xiaonanhai National Geopark, Chongqing
	85	Yangchun Lingxiaoyan National Geopark, Guangdong
2005	86	Mount Taishan National Geopark, Shandong
	87	Dali Mount Cangshan National Geopark, Yunnan
	88	Zhengzhou Yellow River National Geopark, Henan
	89	Mount Tianzhushan National Geopark, Anhui
	90	Jingpohu National Geopark, Heilongjiang
	91	Dehua Mount Shiniushan National Geopark, Fujian
	92	Mount Dabieshan (liu'an) National Geopark, Anhui
	93	Shenzhen Dapeng Peninsula National Geopark, Guangdong
	94	Shehong Silicified Wood National Geopark, Sichuan
	95	Mount Siguniangshan National Geopark, Sichuan
	96	Pingnan Baishuiyang National Geopark, Fujian
	97	Fengkai National Geopark, Guangdong
	98	Fenghuang National Geopark, Hunan
	99	Mount Guanshan National Geopark, Henan
	100	Lincheng National Geopark, Hebei
	101	Mount Yimengshan National Geopark, Shandong
	102	Mount Sanqingshan National Geopark, Jiangxi
	103	Yong'an National Geopark, Fujian
	104	Shennongjia National Geopark, Hubei
	105	Jiuzhi Nianbaoyuze National Geopark, Qinghai
	106	Mount Fengshan Karst National Geopark, Guangxi
	107	Luoninng Shenlingzhai National Geopark, Henan
	108	Wuan National Geopark, Hebei
	109	Fuyun Cocoattuohai National Geopark, Xinjiang
	110	Luoyang Mount Daimeishan National Geopark, Henan

(continued)

Table 10.2 (continued)

Designated year	Number	Geopark name
	111	Yanchuan Yellow River Shequ National Geopark, Shaanxi
	112	Germu Mount Kunlun National Geopark, Qinghai
	113	Mount Huayingshan National Geopark, Sichuan
	114	Mount Changshan Liedao Islands National Geopark, Shandong
	115	Liupanshui Mount Wumengshan National Geopark, Guizhou
	116	Huzhu Mount Beishan National Geopark, Qinghai
	117	Xinyang Jingangtai National Geopark, Henan
	118	Guzhang Red Stone Forest National Geopark, Hunan
	119	Jiangyou National Geopark, Sichuan
	120	Mount Wutai National Geopark, Shanxi
	121	Liuhe National Geopark, Jiangsu
	122	Alashan Desert National Geopark, Inner Mongolia
	123	Luzhai xiangqiao National Geopark, Guangxi
	124	Mount WugongshanNational Geopark, Jiangxi
	125	Dalian Coast National Geopark, Liaoning
	126	Jiufujiang National Geopark, Hunan
	127	Xinggaihu National Geopark, Heilongjiang
	128	Pingtang National Geopark, Guizhou
	129	Zhada Soil Forest National Geopark, Tibet
	130	Benxi National Geopark, Liaoning
	131	Yunyang Longgang National Geopark, Chongqing
	132	Mount Mulanshan National Geopark, Hubei
	133	Huguan Mount Taihangshan Canyon National Geopark, Shanxi
	134	Ningwu Ice Cave National Geopark, Shanxi
	135	Enping Geothermal National Geopark, Guangdong
	136	Yunxian Dinosaur National Geopark, Hubei
	137	Dalian Bingyugu National Geopark, Liaoning
	138	Chongming Yangtse River Delta National Geopark, Shanghai
2009	139	Hongkang National Geopark, Hongkong
	140	Mount Changbaishan Volcano National Geopark, Jilin
	141	Lijiang Yulong Snow Maintain and Glacier National Geopark, Yunnan
	142	Mount Tianshan Tianchi National Geopark, Xinjiang
	143	Mount Wudangshan National Geopark, Hubei
	144	Zhucheng Dinosaur National Geopark, Shandong
	145	Chizhou Mount Jiuhuashan National Geopark, Anhui

(continued)

Table 10.2 (continued)

Designated year	Number	Geopark name
	146	Jiuxiang Canyon and Cave National Geopark, Yunnan
	147	Erlianhaote Dinosaur National Geopark, Inner Mongolia
	148	Kuche Canyon National Geopark, Xinjiang
	149	Liancheng Guanzhaishan National Geopark, Fujian
	150	Qiandongnan Miaoling National Geopark, Guizhou
	151	Lingwu National Geopark, Ningxia
	152	Mount Dabashan National Geopark, Sichuan
	153	Sinan Wujiang Karst National Geopark, Guizhou
	154	Mount Wulongshan National Geopark, Hunan
	155	Hezheng Paleobiologic Fossil National Geopark, Gansu
	156	Dahua Qibainong National Geopark, Guangxi
	157	Mount Guangwushan-Nuoshuihe River National Geopark, Sichuan
	158	Jiangning Mount Tangshan Fangshan National Geopark, Jiangsu
	159	Ningcheng National Geopark, Inner Mongolia
	160	Wansheng National Geopark, Chongqing
	161	Yangbajing National Geopark, Tibet
	162	Shangnan Jinsixia National Geopark, Shaanxi
	163	Guiping National Geopark, Guangxi
	164	Qingzhou National Geopark, Shandong
	165	Xinglong National Geopark, Hebei
	166	Miyun Mount Yunmengshan National Geopark, Beijing
	167	Mount Baiyunshan National Geopark, Fujian
	168	Yangshan National Geopark, Guangdong
	169	Meijiang National Geopark, Hunan
	170	Qian'an-qianxi National Geopark, Hebei
	171	Mount Dabieshan(Huanggang) National Geopark, Hubei
	172	Tianshui Mount Maijishan National Geopark, Gansu
	173	Xiaoqinling National Geopark, Henan
	174	Guide National Geopark, Qinghai
	175	Pinggu Huangsongyu National Geopark, Beijing
	176	Hongqiqu-Mount Linlvshan National Geopark, Henan
	177	Lingchuan Wangmangling National Geopark, Shanxi
	178	Qijiang Wood Fossil-Dinosaur National Geopark, Chongqing

(continued)

Table 10.2 (continued)

Designated year	Number	Geopark name
	179	Yichun xiaoxinganling National Geopark, Heilongjiang
	180	Langao Mount Nangongshan National Geopark, Shaanxi
	181	Qianan Mud Forest National Geopark, Jilin
	182	Fengyang Mount Jiushan National Geopark, Anhui
	183	Datong Volcano Clusters National Geopark, Shanxi
2011	184	Luoping Biota National Geopark, Yunnan
	185	Laiyang Cretaceous National Geopark, Shandong
	186	Turpan Huoyanshan National Geopark, Xinjiang
	187	Wensu Yaanqiu National Geopark, Xinjiang
	188	Luxi Alu National Geopark, Yunnan
	189	Yizhou Shilin National Geopark, Guangxi
	190	Bingling danxia National Geopark, Gansu
	191	Pingshun Tianjishan National Geopark, Shanxi
	192	Xingtai Valley Group National Geopark, Hebei
	193	Zhashui Karst Cave National Geopark, Shaanxi
	194	Pinghe Mount Lingtongshan National Geopark, Fujian
	195	Pinghe Lingtongshan National Geopark, Shanxi
	196	Pingjiang Shiniuzhai National Geopark, Hunan
	197	Zhenghe Fozishan National Geopark, Fujian
	198	Guangde Taijidong National Geopark, Anhui
	199	Pubei Wuhuangshan National Geopark, Guangxi
	200	Yashan National Geopark, Anhui
	201	Zhangye Geopark, Gansu
	202	Yiyuan Mount Lu Geopark, Shandong
	203	Wufeng Geopark, Hubei
	204	Chishui Danxia Geopark, Guizhou
	205	Qinghai Lake National Geopark, Qinghai
	206	Chengde Geopark, Hebei
	207	Fusong Geopark, Jilin
	208	Bayannaoer Geopark, Inner Mongolia
	209	Youyang National Geopark, Chongqing
	210	Ordos Geopark, Inner Mongolia
	211	Ruyang Dinosaur Geopark, Henan
	212	Qingchuan Earthquake Remains Geopark, Sichuan
	213	Xianning Mount Jiugong-Hot Spring Geopark, Hubei
	214	Mount Yao Geopark, Henan
	215	Yaozhou Zhaojin Danxia Geopark, Shaanxi

(continued)

Table 10.2 (continued)

Designated year	Number	Geopark name
	216	Mianzhu Qingping-Hanwang Geopark, Sichuan
	217	Maqin Mount Anyemaqen Geopark, Qinghai
	218	Mount Dawei Geopark, Hunan
	219	Mount Phoenix Geopark, Heilongjiang
2014	220	Enshi Tenglongdong Grand Canyon Geopark, Hubei
	221	Du'an Underground River Geopark, Guangxi
	222	Tongdao Mount Wanfoshan Geopark, Hunan
	223	Changyang Qingjiang Geopark, Hubei
	224	Changle Geopark, Shandong
	225	Jinzhou Geopark, Liaoning
	226	Qingshuihe Laoniuwan Goepark, Inner Mongolia
	227	Anhua Xuefenghu Geopark, Hunan
	228	Yulin Fossil Geopark, Shanxi
	229	Lingbi Mount Qingyunshan Geopark, Anhui
	230	Siping Geopark, Jilin
	231	Qingliu Hot Spring Geopark, Fujian
	232	Sanming Jiaoye Geopark, Fujian
	233	Luocheng Geopark, Guangxi
	234	Shicheng Geopark, Jiangxi
	235	Dangchang Guan'egou Geopark, Gansu
	236	Siziwang Geopark, Inner Mongolia
	237	Fanchang Mount Marenshan Geopark, Anhui
	238	Lintan Yeliguan Geopark, Gansu
	239	Lianyungang Mount Huaguoshan Geopark, Jiangsu
	240	Shankou Geopark, Heilongjiang
	241	Huludao Longtan Valley Geopark, Liaoning

10.7 Future Prospects for Chinese National Geoparks

Tourism demand and conservation of geosites are the driving forces for the development of geoparks. The geological conditions and variety of geolandscape resources in China form the basis for the establishment of geoparks.

10.7.1 Rapid Development of China's Tourism Increases the Demand for Geoparks

Geoparks are an important tourist destination whose number largely depends on the demands of the tourist market. The tourist market relies on the number of tourists and their opinions about geoparks. China has substantial room for growth in tourist numbers. In developed countries, every citizen travels around six times per year, with an average of eight times in any individual country. In China, the annual number of trips per citizen is less than two, but even at this low rate, many attractions will be overcrowded in the tourist season. China needs to build more tourist destinations to meet tourism demand. From this perspective, natural parks that disseminate scientific knowledge are becoming more and more popular. As natural parks with substantial scientific values, excellent natural conditions and wonderful landscapes, geoparks have become the most popular natural park, which highlights a bright future for them (Chen 2006).

10.7.2 Preserving GeoHeritage Demands Geoparks

Geosites are records formed over the evolutionary history of the earth, providing irreversible evidence of earth's evolution, and the origins of life and species evolution, and accordingly are deserving of rigorous protection. The establishment of geoparks is the best solution to preserving our geoheritage, which provides a rationale for the development of more geoparks.

10.7.3 Resources to Develop Geoparks in China

China has a vast territory, with a similar land area to the whole of Europe. It is located at the confluence of several tectonic plates, which creates complex geological conditions. Geolandscapes in China are classified according to various types, quality and distribution. Almost all types of geolandscapes throughout the world can be found in China, particularly on the unique Tibetan Plateau and boundless Karst land loess landscapes. China shows great resource potential to develop geoparks of various types and levels.

10.7.4 Prospects for Chinese Geoparks

Based on the rapid increase in tourism requirements for geopark construction and geoheritage protection, and particularly the resources available to develop geoparks, an additional 50 global geoparks, 300 national geoparks, 500–750 provincial geoparks and many county geoparks are predicted to be developed by the middle of this century. This will result in an orderly and evenly distributed geopark system with all kinds of geoheritages in China (Chen 2008).

10.8 European and Global Geoparks

10.8.1 A Brief History and the Present Situation
of European Geopark Development

Europe was an early initiator of the establishment of continent-wide geoparks. At the 30th International Geological Congress in Beijing in August 1996, Professors Guy Martini (France) and Nickolaos Zouros (Greece) proposed the initiative "to establish the European geopark", which was supported by the European Union. In June 2000, the European Geoparks Network (EGN) was established (Eder and Patzak 2001). At the first European Geopark Congress in Spain in November 2000, the first four European geoparks were announced. Up until 2012, a total of 50 geoparks representing 17 nations had joined the EGN. A list of European geoparks is provided in Table 10.3.

10.8.2 Brief History and Present State of Global Geopark
Development

UNESCO passed the *Convention Concerning the Protection of the World Cultural and Natural Heritage* in 1972 and initiated the World Heritage list. However, only a small number of geological heritage sites were present on the list before 1995. Most of the world's precious geoheritage sites are not effectively protected, causing great concern among geologists. At the 30th International Geological Congress in Beijing in August 1996, the UNESCO Division of Earth-science and the International Union of Geological Sciences together proposed the "build Global Geopark" initiative in order to protect more of our global geological heritage, and the term "Geopark" was suggested. The initiative immediately attracted more than 30 supporters including China. At the 29th plenary session of UNESCO in November the following year, the decision to "create a global network of geoheritage with unique geological characteristics" was announced.

The goal of promoting the establishment of geoparks worldwide was more effective protection of geoheritage. At the 156th Executive Board meeting of UNESCO in April 1999, the *UNESCO Geopark-Programme* was officially launched and the UNESCO Network of Geoparks (or GGN) was established (UNESCO 1999a, b). In 2000, the formal establishment of Chinese national geoparks and European geoparks promoted the work of GGN. The earth-sciences division of UNESCO promoted the development of GGN to all of the delegates at the 31st International Geological Congress in Brazil in August 2000 and formally promulgated the *Operational Guidelines for the Network of Geoparks under*

Table 10.3 List of global geoparks network members

	Geopark name	Year designated	Country
1	Nature Park Eisenwurzen	2004	Austria
2	Huangshan Geopark		China
3	Wudalianchi Geopark		
4	Lushan Geopark		
5	Yuntaishan Geopark		
6	Songshan Geopark		
7	Zhangjiajie Sandstone Peak Forest Geopark		
8	Danxiashan Geopark		
9	Stone Forest Geopark		
10	Reserve Géologique de Haute Provence		France
11	Park Naturel Régional du Luberon		
12	Nature park Terra Vita		Germany
13	Geopark Bergstrasse–Odenwald		
14	Vulkaneifel Geopark		
15	Petrified Forest of Lesvos		Greece
16	Psiloritis Natural Park		
17	Marble Arch Caves and Cuilcagh Mountain Park		Ireland, Republic of/Northern Ireland
18	Copper Coast Geopark		Republic of Ireland
19	Madonie Natural Park		Italy
20	Maestrazgo Cultural Park		Spain
21	North Pennines AONB Geopark		UK
22	Hexigten Geopark	2005	China
23	Yandangshan Geopark		
24	Taining Geopark		
25	Xingwen Geopark		
26	Bohemian Paradise Geopark		Czech Republic
27	Geopark Harz Braunschweiger Land Ostfalen		Germany
28	Geopark Swabian Albs		
29	Parco del Beigua		Italy
30	Hateg Country Dinosaur Geopark		Rumania
31	North West Highlands—Scotland		UK
32	Forest Fawr Geopark—Wales		

(continued)

Table 10.3 (continued)

	Geopark name	Year designated	Country
33	Araripe Geopark	2006	Brazil
34	Taishan Geopark		China
35	Wangwushan-Daimeishan Geopark		
36	Funiushan Geopark		
37	Leiqiong Geopark		
38	Fangshan Geopark		
39	Jingpohu Geopark		
40	Gea- Norvegica Geopark		Norway
41	Naturtejo Geopark		Portugal
42	Sobrarbe Geopark		Spain
43	Subeticas Geopark		
44	Cabo de Gata Natural Park		
45	Papuk Geopark	2007	Croatia
46	Geological and Mining Park of Sardinia		Italy
47	Langkawi Island Geopark		Malaysia
48	English Riviera Geopark		UK
49	Longhushan Geopark		China
50	Zigong Geopark		
51	Adamello Brenta Geopark		Italy
52	Rocca Di Cerere Geopark		
53	Alxa Desert Geopark	2009	China
54	Zhongnanshan Geopark		
55	Chelmos-Vouraikos Geopark		Greece
56	Toya Caldera and Usu Volcano Geopark		Japan
57	Unzen Volcanic Area Geopark		
58	Itoigawa Geopark		
59	Arouca Geopark		Portugal
60	Geo Mon Geopark—Wales		UK
61	Shetland Geopark		
62	Stonehammer Geopark	2010	Canada
63	Leye-Fengshan Geopark		China
64	Ningde Geopark		
65	Rokua Geopark		Finland
66	Vikos—Aoos Geopark		Greece
67	Novohrad-Nograd geopark		Hungary-Slovakia
68	Parco Nazionale del Cilento e Vallo di Diano Geopark		Italy
69	Tuscan Mining Park		
70	San'in Kaigan Geopark		Japan

(continued)

Table 10.3 (continued)

	Geopark name	Year designated	Country
71	Jeju Island Geopark		Korea
72	Magma Geopark		Norway
73	Basque Coast Geopark		Spain
74	Dong Van Karst Plateau Geopark		Vietnam
75	Tianzhushan Geopark	2011	China
76	Hongkong Geopark		
77	Bauges Geopark		France
78	Geopark Muskau Arch		Germany/Poland
79	Katla Geopark		Iceland
80	Burren and Cliffs of Moher Geopark		Ireland, Republic of/Northern Ireland
81	Apuan Alps Geopark		Italy
82	Muroto Geopark		Japan
83	Sierra Norte di Sevilla, Andalusia		Spain
84	Villuercas Ibores Jara Geopark		
85	Carnic Alps Geopark	2012	Austria
86	Sanqingshan Geopark		China
87	Chablais Geopark		France
88	Bakony-Balaton Geopark		Hungary
89	Batur Geopark		Indonesia
90	Central Catalunya Geopark		Spain
91	Shennongjia Geopark	2013	China
92	Yanqing Geopark		
93	Sesia—Val Grande Geopark		Italy
94	Oki island Geopark		Japan
95	Hondsrug Geopark		Netherlands
96	Azores Geopark		Portugal
97	Idrija Geopark		Slovenia
98	Karavanke/Karawanken		Slovenia and Austria
99	Kula Volcanic Geopark		Turkey
100	Grutas del Palacio Geopark		Uruguay
101	Ore of the Alps Geopark		Austria
102	Tumbler Ridge Geopark		Canada
103	Mount Kunlun Geopark		China
104	Dali Mount Cangshan Geopark		China
105	Odsherred Geopark	2014	Denmark
106	Mount d'Ardeche Geopark		France
107	Aso Geopark		Japan

(continued)

Table 10.3   (continued)

	Geopark name	Year designated	Country
108	M'Goun Global Geopark		Morocco
109	Lands of Knights Geopark		Portugal
110	El Hierro Global Geopark		Spain
111	Molina and AltoTajo Geopark		Spain

As of October 2014, 111 geoparks in 32 member states were members of the Global Geoparks Network assisted by UNESCO

UNESCO's Patronage declaring the way to nominate a global geopark and began to accept applications. In June 2004, at the first Beijing Global Geopark Congress, both the *Global Geopark Congress Charter* and the *Geo-heritage Protection: Beijing Declaration* was passed. Also, the first 25 global geoparks, including eight Chinese geoparks (e.g. Mount Huangshan Global Geopark) were announced. The Global Geopark Network Office was formally founded in Beijing, initially under the management of the Chinese MLR, and from October 2010, the Chinese Academy of Geo-sciences. As of October 2013, 111 geoparks from 32 countries had joined the GGN.

10.8.3 Development History and Present State of Global Geoparks in China

Since establishment of the first national geoparks in 2000, China has made tireless efforts in promoting the establishment of global geoparks to support the UNESCO initiative. After UNESCO announced the *Global Geopark Network Guide* in 2002, China developed processes to establish global geoparks and nominated eight geoparks (including Mount Huangshan National Geopark) for global geopark status. In 2004, following site visits by experts from the UNESCO, the Global Geopark Committee voted in favour of all eight nominated parks becoming China's first global geoparks.

As of 2014, there were 111 global geoparks, 31 of which are in China, which is the country with the most global geoparks: Huangshan Geopark, Wudalianchi Geopark, Lushan Geopark, Yuntaishan Geopark, Songshan Geopark, Zhangjiajie Sandstone Peak Forest Geopark, Danxiashan Geopark, Shilin Geopark, Hexigten Geopark, Yandangshan Geopark, Taining Geopark, Xingwen Geopark, Taishan Geopark, Wangwushan-Daimeishan Geopark, Funiushan Geopark, Leiqiong Geopark, Fangshan Geopark, Jingpohu Geopark, Longhushan Geopark, Zigong Geopark, Alxa Desert Geopark, Zhongnanshan Geopark, Leye-Fengshan Geopark, Ningde Geopark, Tianzhushan Geopark, Hongkong Geopark, Sanqingshan Geopark, Shennongjia Geopark, Yanqing Geopark, Mount Kunlun Geopark and Dali Mount Cangshan Geopark. As the former director of UNESCO's Division of

Earth-sciences, Wolfgang Eder, said: "China has played a pioneering role in promoting the establishment of geoparks". China has made a significant contribution to the creation of global geoparks in UNESCO network.

10.8.4 The Future Prospects of Global Geoparks

The rapid development of global geoparks suggests that they are in demand by the public, and are gaining increasing support from governments because of their advanced concepts. Global geoparks are similar to world heritage and wetlands, but the main purpose of the latter was protection rather than land use, whereas global geoparks aim to conserve and protect geoheritage and natural environments. Geoparks also aim to disseminate knowledge of earth-science to the public, and especially to activate geotourism that boosts employment for local communities around the park and facilitates local economic development, thus largely improving people's quality of life and enriching their spiritual life. In this sense, the geopark idea is comprehensive, because it ties together protection and use and promotes sustainable development of local economies, cultures and society, which lays a solid foundation for the development of geoparks. In 2000, UNESCO's final goal for the number of global geoparks was 500 (Eder 1999), but this number may reach 800–1000 by the end of the twenty-first century. All countries will construct more national, state and county geoparks. Thus, the future of the geopark is very bright.

Geoparks were a new idea, arising from the theory and practice of tourism geoscience and growing up under its guidance. The construction of Chinese geoparks promoted the establishment and development of global geoparks. The vigorous growth of the geopark will lead to the development and improvement of tourism geoscience. They promote one another and improve together, confirming a bright future for the geopark concept.

References

Chen, A. (1996). 10 years of tourism earth-science—in Commemoration of the 10th anniversary of the foundation of tourism earth-science institute. *Tourism Tribune, 1*: 58–61, 79.

Chen, A. (1998). The born of tourism earth-science and its tasks: for tourism tribune tourism earth-science album. *Tourism Tribune*, (suppl), 1–2.

Chen, A. (2006). Twenty years of initiating and innovating tourism—in commemoration of the 20th anniversary of the founding of Tourism Earth-Science Research Association. *Tourism tribune, 4*(21), 71–77.

Chen, A. (2008). Development situation, problems and solutions of Chinese National Geopark. *Tourism geoscience and geopark construct—the 14th collected works of tourism geoscience.* Beijing: China Forestry Press.

Chen, A., Jiang, J., & Li, M. (2007). Development situation and prospects of Chinese National Geopark. *Green book of China's tourism.* Beijing: Social Sciences Academic Press.

Eder, W. (1999). "UNESCO Geoparks"—A new initiative for protection and sustainable develop-
ment of the Earth's heritage. *Neues Jahrbuch Fur Geologie Und Palaontologie-Abhandlungen,*
214(1/2), 353–358.
Eder, W., & Patzak, M. (2001). *Geological heritage of UNESCO European Geoparks network.*
Spain: European Geoparks Magazine.
UNESCO. (1999a). Operational Guidelines for UNESCO Geoparks.
UNESCO. (1999b). Geopark Nomination form. 02.

Chapter 11
Prospects for Tourism Earth-science and Geotourism

An Introduction to Tourism Earth-science (Chinese Version, 1991) is a summary of the research findings of the Chinese Academy of Tourism Earth-science and Geopark Research since its establishment in 2009, as well as that of China's tourism earth-science circle. It marks a new stage in the development of China's tourism earth-science research. However, as mentioned in Chap. 1 of that book, tourism earth-science only recently became a discipline in its own right in China and worldwide. It was not until 1985 that the term was used in China. As an emerging discipline, tourism earth-science is still under development. Therefore, it is timely to assess its future and make some suggestions regarding its ongoing development.

11.1 Increase in Social Demand for Tourism Earth-science and Geotourism

The emergence of a discipline entails social demand and independent research objects, methods and a theoretical basis. Tourism earth-science is a discipline applying earth-science theory and other scientific theories and methods (e.g. those of aesthetics, landscape science, environmental science and tourism science) to research earth-science problems relating to tourism, thereby serving tourism. In China, it came into being to meet the needs of growing tourism. Therefore, its prospects are associated with those of tourism. As early in 1996, the World Travel and Tourism Council (WTTC) pointed out that tourism was the largest global industry measured by all indicators (Liang 2013). According to statistics from the Japanese National Tourism Office, the output value of world tourism reached US$1606 billion in 2007, more than that of the telecommunication (~US$1515 billion), chemical (US$1484 billion) and auto (US$1183 billion) industries; making tourism largest global industry. Based on projections

© Springer-Verlag Berlin Heidelberg and Science Press Ltd. 2015
A. Chen et al., *The Principles of Geotourism*, Springer Geography,
DOI 10.1007/978-3-662-46697-1_11

by the World Tourism Organization (WTO), international visitor numbers will increase to 1.62 billion and tourism income will reach US$2000 billion, indicating a promising future for the world tourism industry. Although China is currently placed second in the world in terms of its economy, its tourism is rapidly developing. According to the Chinese Economic and Social Development Statistical Communique of 2013, numbers of national tourists in China reached 3.26 billion in 2013, with 129.08 million incoming tourists; domestic tourism income reached 2627.6 billion RMB, with US$51.7 billion of international tourism income and 98.19 million outbound tourists. According to a report from the Head of the Chinese National Tourism Office, (Shao Qiwei) in early 2014, annual domestic tourist numbers will increase to 3.58 billion, with 56.8 million inbound overnight tourists and 114 million outbound tourists; creating 3190 billion RMB of total tourism income. This makes China a huge player in the global tourism industry. In summary, tourism earth-science is certain to usher in a new era of rapid development together with booming tourism in China.

11.1.1 A New Phase of Tourism Development

Tourists, tourism resources and tourism service facilities are the three major components of tourism, all of which are essential to the development of tourism. All are related to earth-science, with tourism resources most closely linked to earth-science. It was not until the late 1970s that China's tourism began to develop extensively. In its initial stages, the principal problem was insufficient tourism service facilities. At that time, efforts focused on building hotels and airports, but such projects were not urgent with respect to tourism earth-science. However, with the passage of time, service facilities have been improved and the number of tourists has increased significantly so that some existing scenic areas and spots are saturated or super-saturated with tourists. Moreover, insufficient variety and unbalanced distribution of scenic areas and sports are aggravating their contradiction with tourism resources. Beijing, for example, has diverse tourism resources, but its existing scenic sites are overcrowded. Moreover, local people are losing interest in the existing tourist spots and are desperate for new tourism resources. Sites such as Shidu, Yesanpo, Longqing Ravine, Shihua Cavern, Mount Yunmeng, Jingdong Big Karst Cave and Kangxi Grassland have been discovered in recent years, and the same period has seen the establishment of five national geoparks (including Shihua Cavern), two UNESCO world geoparks (Fangshan and Yanqing) and two national mining parks (Miyun and Huairou). These have added to the diversity of tourism resources and eased the shortage of tourism resources and are well received by tourists.

Currently, central and local governments are formulating tourism development plans (including geopark development plans) and require information on the status of tourism resources. However, existing information about the overall status of tourism resources is inadequate, leading to difficulties or errors in tourism planning. Local governments are now aware of the importance of surveys and have begun to develop general surveys of tourism resources. During such surveys, and through evaluation and planning of tourism resources, many problems related to tourism earth-science may be revealed, which will require the attention of tourism earth-science and geotourism workers. This provides opportunities for the development of tourism earth-science and ushers in a new era of prosperity for tourism earth-science.

11.1.2 The Urgent Need for Compiling Course Books on Tourism Earth-science

To meet the needs of tourism development, tourism education is developing: tourism earth-science has become a compulsory course in both tourism and earth-science colleges; departments of geography of comprehensive and normal universities have developed courses relating to tourism earth-science (such as tourism geography and preliminary tourism earth-science); most higher and middle geological institutions have set courses in tourism geology, including some that are planning to provide tourism geology or tourism earth-science. Therefore, text books to support these courses are urgently required. However, development of a comprehensive curriculum and associated course books entails extensive scientific research and analysis of substantial data. The development of tourism education promotes the processing and theoretical summarizing of practical materials for tourism earth-science and advances the maturation and rapid development of tourism earth-science and geotourism. Most existing course books and monographs in tourism earth-science are related to tourism geography, and some collections of tourism earth-science papers and common readers have also been published. All of these are important components of tourism earth-science. *An Introduction to Tourism Earth-science* is a comprehensive theoretical work that can be used as a course book or reference book for tourism education and its related courses.

11.1.3 The Need to Increase the Earth-science Knowledge of Tourist Guides

The development of cultural education has contributed to the increasing cultural literacy of the whole nation and is elevating tourism to a higher level. Tour guide explanations centred on myths and legends are falling short of the needs of tourists. Even tourists with minimal schooling are losing interest in routine explanations about caves or landscapes, as they connect with more and more scenic areas. They are eager to gain scientific knowledge about how such scenic wonders were formed. Therefore, guides are generally required to provide earth-science explanations. Geoscience-based and scientific expedition tourism, which differs from ordinary tourism, is increasing. It has already been recommended to the China National Tourism Office that tourism earth-science accreditation is required for the issue of a tour guide certificate. To that end, *Tourism Earth-science coursebook for Tour Guides* is required. In recent years, as China increasingly opens its doors to the outside world, it has hosted many international earth-science conferences, which typically include post-conference geological tours. As the chair of the Geo-tour Steering Committee at the 30th International Geological Congress held in Beijing in 1996, Chen Anze organized over 120 geotours before, during and after the conference; attended by over 1300 of the 6000 congress registrants. A series of six mega-volumes of geotour guides in English were compiled and published as an important tourism earth-science work.

The above-mentioned issues present a challenge for tourism earth-science and will provide the momentum for the rapid development of tourism earth-science. An emerging marginal discipline will be revealed to tourists.

11.2 Evaluation of Tourism Resources Will Become a Key Research Topic

In years to come, many areas will need to be set aside (including geoparks and mining parks) in order to serve tourism demands (Chen and Jiang 2004); accordingly, tourism resources will be the principal contradiction for the development of tourism. Finding a solution to this contradiction involves four aspects: (1) search, (2) evaluation, (3) development and utilization and (4) protection of tourism resources—of which the first two are key problems. Many aspects must be considered in the search of tourism resources, including classification, formation causes, formation and rules of distribution of resources. Finding a new tourism resource requires substantial work, but evaluating such resources seems more difficult. As most tourism resources are distributed on the earth's surface (although some are underground, such as underground caves, fossils of ancient lives and tombs), they will eventually be found with sufficient effort. However, it is more difficult to establish the value of tourism resources. Although there are more evaluation elements, there is no recognized evaluation standard, which makes it very difficult to accurately evaluate tourism resources. Strictly, tourism resources that have not been scientifically evaluated should not be developed, because it is difficult and risky. Therefore, an urgent research task for tourism earth-science is to promptly compile a full array of evaluation standards for varied tourism resources.

11.3 Research Teams in Tourism Earth-science Will Continue Expanding

The revitalization of tourism has encouraged experts and scholars from all disciplines to bring to tourism the strengths of their respective disciplines. Experts in various disciplines in earth-sciences circles also join research teams in tourism. Tourism earth-science workers find it necessary to establish a horizontal academic group to facilitate academic exchanges, learn from tourist and other academic circles and better serve tourism, hence the appearance of the Chinese Academy of Tourism earth-science and Geopark Research. It currently has more than 500 members in China and has established local tourism research collaborations in more than 10 provinces. Its activities in recent years have played a very important role in uniting tourism earth-science workers and boosting research in tourism earth-science and have won widespread recognition from tourist and earth-sciences circles. Local tourism earth-science research associations are expected to appear more quickly. The activities of the organization will shift their focus from nationwide to local, and their contents will turn from comprehensive to thematic. The organization is a mass academic group horizontally integrating people from earth-sciences, tourist and other academic circles, aiming to discuss relationships between earth-science and tourism, and how to serve China's tourism and promote the progress and development of tourism earth-science in China. The integration of disciplines is an important way to upgrade each respective discipline and develop the new combined discipline. All tourism earth-science workers should promote the spirit of cooperation and learn from each other so as to advance the emerging marginal discipline of tourism earth-science. As tourism earth-science develops its theoretical disciplinary system, it will find wider applications in China's tourism. The development of tourism promotes the establishment and development of tourism earth-science as a discipline, and reciprocally, the development of tourism earth-science will boost the prosperity of tourism. The two promote each other and develop side by side, creating a bright future for tourism earth-science.

11.4   Strategic Conception of Tourism Earth-science and Geotourism Over the Next Ten Years (2015–2025)

"Things would go on smoothly if well planned and prepared; and they would go to a failure if bad prepared and planned". This is a famous quotation from *the Book of Rites*. Approaching its 30th anniversary and entering the fourth decade of development, a strategic plan is being developed for tourism earth-science so that new achievements will be made towards its sustainability. This plan will focus on factors that raise the spirit of tourism earth-science, finalize its scientific system and maximize its influence at home and abroad, in order that it can make a larger contribution to China's tourism industry and geopark causes. The strategic plan encompasses four aspects in relation to tourism earth-science: organizational mission, targeting systems building, development strategies and implementation schemes for this development.

11.4.1   Organizational Mission of Tourism Earth-science

1. The central task of tourism earth-science is to ensure it remains visible in China's tourism industry, and to transmit knowledge of the discipline to the general public through travel and tourism. Geotourism will be promoted as

a new category of tourism, functioning as the core of nature-based tourism over the next decade. The theories and methodologies would be applied self-consciously in the tourism industry to create tourism products, carry out geotourism activities and enhance the persistence of tourism earth-science in the tourism industry.

2. Tourism earth-science must make sure that the discipline influences how job vacancies are filled in geoparks. A geopark is a child of tourism earth-science theories and practices. The directory role of tourism earth-science will be enhanced by the establishment, building, management and site promotion of geoparks. Geopark workers must be made aware that tourism earth-science theories are required instruments for directing park building and management (Chen 2013a).

3. Tourism earth-science must influence reference books used in tourism. Works such as *A Grand Tourism Geoscience Dictionary* (Chen 2013b) and *Library of China's National Geoparks* are compiled for the purpose of tourism earth-science; accordingly, they should be necessary reference books for geopark managers, tour guides, tourism planners, geopark planners, exhibit designers, scientific researchers, science promotion editors, science promotion activity organizers and tourism, geology or geography college professors and students.

11.4.2 The Targeting Systems Building in Tourism Earth-science

The targeting systems include the discipline theoretical system, organizational system, research body system, teaching and learning system, training system, achievements evaluation system and discipline promotion system. These can be described as follows:

1. Discipline theoretical system building. Tourism earth-science monographs will be compiled for publication so that the theoretical construction of the discipline is enhanced over the next decade. Such monographs include the revised versions of *An Introduction to Tourism Earth-science, General Course Readings for Tourism Earth-science, Tourism Earth-science Manuals for Tour Guides, English Version of An Introduction to Tourism Earth-science, Revised Version of Grand Dictionary of Tourism Earth-science, An Introduction to Landscape Resources of Tourism Earth-science, Aesthetic Appreciation on Landscape Resources of Tourism Earth-science, Studies on National Geoparks, History of China's National Geoparks, History of UNESCO World Geoparks, Library of China's National Geoparks, Serial Science Promotion Oriented Tour Guide Maps of China's National Geoparks, History of China's Tourism Earth-science,* and *Serial Symposia of Tourism Earth-science.*

2. Organizational system building. Networking of the Chinese Academy of Tourism earth-science and Geopark Research Branch of the Chinese Geological Society must be achieved to ensure regional branch offices are established in every

provincial administration. A mechanism of increasing group and individual membership should be created (or resumed) to ensure that individual member numbers reach 2000 and group members, 50–100. As a group member of the Chinese Geological Society, its Tourism earth-science and Geopark Branch has established a close relationship with the World Geotourism Organization to ensure that global geotourism is encouraged and promoted.

3. Research body system building. The Chinese Academy of Tourism Earth-science and Geopark Research should be established as the home base to create a further 20–30 Tourism earth-science Research Centers (including those in Hong Kong and Taiwan).

4. Teaching and learning system building. Tourism earth-science and geotourism should be introduced as a compulsory or selected teaching course in 10–20 higher education institutions, creating masters and doctorate programmes to produce 1000–2000 masters and doctors of tourism earth-science and geotourism.

5. Training system building. All levels of tourism earth-science research associations and research centres should be identified as intellectual platforms through which geopark managers, earth-science tour guides, and professional designers and planners can be trained and developed. It is recommended that the mastery of certain knowledge of tourism earth-science be essential under a national policy regulating qualifications of tour guides, and a prerequisite for geopark management staff.

6. Research achievements evaluation system building. A Tourism earth-science Research Fund should be created via associated parties in order to support tourism earth-science research programs. Tourism earth-science evaluation organization should be based at a Tourism earth-science Research Branch to conduct evaluation activities, publicize excellent projects and reward researchers with prizes as encouragement.

7. Discipline promotion system building. A professional group of tourism earth-science promoters should be created from the Chinese Academy of Tourism Earth-science and Geopark Research members, transmitting the essence of the discipline into higher education institutions and more broadly, to international communities.

11.4.3 The Development Strategy for Tourism Earth-science and Geotourism

1. Establish geoparks as the central platform on which geotourism will develop as an important tourism programme in China.

2. Directed by the theories of tourism earth-science, geopark ideas should be carried through into the development of geoparks.

3. Set practice, theory and repractice to improve the tourism earth-science discipline system, to enhance tourism earth-science to a higher level at which China's tourism will follow the road of science-based tourism.

11.4.4 The Implementation Schemes of Tourism
Earth-science Development

1. Cooperate with the World Geotourism Organization (registered in Canada in February 2014). The World Geotourism Organization will be formally set up declared in China in 2015; the secretariat will be established at the Chinese Academy of Geological Sciences; the World Geotourism Forum will occur at the Wanshan Geotourism Culture Industrial Park of Yingyang, in Zhengzhou, Henan. China will become the centre of world geotourism and will highlight the contributions of tourism earth-science to the world.
2. Efforts will be made to launch a "Year of China's Geotourism" to ensure that geotourism develops as a major category of tourism in China.
3. Maintain close ties with tourism earth-science experts and lovers. Organizational, discipline and development strategy superiority will be used to realise all the targeting systems of the tourism earth-science organizational mission.
4. In order to build a solid economic base to support the realization of the tourism earth-science targeting system, efforts will be made to transform working and research achievements into policy support from the government, and other sources of funding support, and finally create the Tourism earth-science Research Fund.

The above-mentioned ideas form a strategic concept for developing tourism earth-science. The author hopes that colleagues in tourism earth-science will offer rectifying and complementary ideas so that plans regarding development of tourism earth-science over the next 10 years are more systematic and complete. It is anticipated that tourism earth-science colleagues, especially those from the younger generation, will both shoulder the burden and inherit the tradition of lifting the spirit of tourism earth-science; further fulfil and complete the discipline's systems; establish tourism earth-science in higher education and scientific research institutions; and transmit knowledge of tourism earth-science to other parts of the world. With the rise of geoparks in 21st century, a global tide of geotourism is rapidly developing. The English version is designed to meet this demand. This is done to provide colleagues in the international community with a fuller understanding of tourism earth-science as created in China and encourage them to support the development of tourism earth-science and geotourism. This will empower tourism earth-science to direct the development of geoparks and conduct global geotourism. This is particularly aimed at the Sixth UNESCO World Geopark Congress held in Canada in September 2014 and the World Geotourism Congress to be held in Zhengzhou, China in 2015. The author is expecting that tourism earth-science will contribute more to the world geotourism cause and to the sustainable development of UNESCO world geoparks.

References

Chen, A. (2013a). On National Geopark. *Tourism earth-science and geopark studies: Chen Anze selected works* (pp. 159–168). Beijing: Science Press.

Chen, A. (2013b). *A Grand tourism earth-science dictionary* (pp. 159–168). Beijing: Science Press.

Chen, A., & Jiang, J. (2004). *China's National Geopark development: Current situation and future prospects—Tourism Blue Book*. Beijing: Social Science Literature Press.

Liang, C. (2013). Northeast Asia sustainable tourism development and international tourism cooperation (Northeast Finance University Doctoral Dissertation).

Postscript

To ensure a better understanding of the concepts of geotourism and tourism earth-science, it is necessary to summarize the previous chapters into the following points:

1. Applying earth-science in developing and promoting tourism is the direct result of the economic revolution of China occurred in the late 1970s. Earth scientists in China have followed the trend of changes and drastic surge in demand for tourism products and services. Since then, a new academic discipline has appeared in China. It has integrated tourism with earth-science and has been academically called 'tourism earth-science' by Chinese earth scientists. Throughout its 30 years of development, tourism earth-science has assisted the healthy growth of China's tourism and forms the breeding ground for geoparks in the country. It has performed as a fundamental science with practical uses in developing tourism products. Tourism earth-science and geoparks are products of earth-science. They have great impacts on tourism and landscape architecture development in China. This Chinese founded academic discipline provides a future direction for science advancement in China.

2. Science has no boundary. Knowledge should be shared with all people. English is an international language. It is also a tool for international communication. Therefore, the translation of the original Chinese version of this book into English is essential to serve the purpose of disseminating the messages contained in the book on a global basis. By having Springer as the publisher of its English edition, the spreading of these messages could be even more extensive. It is hoped that the book would arouse the interest of international scholars, policy makers, teachers, planners, park and tourism managers and operators. By rallying the support of these people, tourism earth-science can then be developed further in the other parts of the world.

3. This book contains 12 chapters. It covers the history of tourism earth-science, its research topics, methodology, application, functions, nature tourism

© Springer-Verlag Berlin Heidelberg and Science Press Ltd. 2015
A. Chen et al., *The Principles of Geotourism*, Springer Geography,
DOI 10.1007/978-3-662-46697-1

resources, ecology and cultural landscapes, planning, protection, geoparks and discussion on the future of tourism earth-science. All earth-science studies are closely associated with tourism and it is only through tourism earth-science that the application of earth-science can be adequately expressed. The term tourism earth-science was being used under such circumstances in 1985. The development of tourism in the twentieth century provides growth of alternative tourism such as ecotourism, forest tourism, cultural tourism, industry tourism, agriculture tourism and geotourism. They are forms of tourism activities instead of academic disciplines. Geotourism is associated with tourism earth-science and is well known in the tourism industry as a tourism activity with focus on geology and natural landscape. Because this the book has been named 'The Principles of Geotourism', after reading this book, the readers should have a better understanding of the differences between geotourism and tourism earth-science.

4. For the maturity and integrity of an academic discipline, it should have extensive support from scholars and this may take decades or even a hundred years to reach that stage. The history of tourism earth-science is still relatively short. It is a concept and a set of principles developed by Chinese scholars through many years of practical field experiences. With limited exchange with other international scholars, it is bound to have plenty of room for improvement. The authors welcome any comments from the readers for upgrading the contents of future editions. It is hoped that this book will arouse the interest of earth-scientists and tourism professionals on the development of this new academic discipline and encourage research in its academic structure and applicability. This will certainly provide earth-scientists with a more significant role to play in the rapidly growing global tourism industry.

Printed in the United States
By Bookmasters

Printed in the United States
By Bookmasters